# Practical
# Wire Antennas

### EFFECTIVE HF DESIGNS
### FOR THE RADIO AMATEUR

John D Heys, G3BDQ

Radio Society of Great Britain

Published by the Radio Society of Great Britain, Cranborne Road, Potters Bar, Herts EN6 3JE.

ISBN 0 900612 87 8

Typography by S. Clark, Radio Society of Great Britain.
Cover Design by Linda Penny, Radio Society of Great Britain.
Printed in Great Britain by Bath Press Ltd., Lower Bristol Road, Bath. BA2 3BL.

# Contents

# Preface

My introduction to the world of antennas happened when I was just 11 years old. I was sitting on the hearth rug and playing with my Christmas present, a small induction or 'shocking coil', when one of its 'handles' accidentally touched the brass fender which ran in front of the fireplace. The family broadcast receiver happened to be switched on, and from its speaker came a loud rasping noise! This chance experience, together with a natural innate curiosity, initiated a lifelong enthusiasm for amateur radio. In 1939, and when still at school, I made a vertically polarised two-element wire beam, which for several summer weekends was fruitlessly pointed towards Mt Snowdon in an attempt to receive the 56MHz signals that were being transmitted from that lofty site.

Wire antennas have always fascinated me, and all through those years that have raced by since I was first licensed in 1946 I have put up and air tested hundreds of them. On the HF bands I have only ever used wire antennas and have often discovered that they can be just as effective as many of the commercially manufactured tribanders that turn atop their lofty and costly towers.

This book has been written by a non-mathematician whose knowledge of this subject has never extended beyond the Cambridge School Certificate syllabus! It is aimed towards anyone who is capable of passing the Radio Amateurs' Examination, and the range of antennas described and illustrated are easy to set up and use successfully. There are additional data which will allow experiments and tests with versions that are cut for other bands or designed to fit into difficult locations. The simplified and, it is hoped, easily understood antenna theory is an attempt to allow the newest recruit to amateur radio to learn something about how simple wire radiators work at HF.

Technical progress has been so rapid over the last 20 years that the home construction of 'state-of-art' HF transceivers lies beyond the capabilities of most radio amateurs. We are therefore obliged to invest in one of the many 'black boxes' that are available. One of the few remaining fields for experiment is 'do-it-yourself' antenna construction, and for just a few pounds of expenditure on wire, insulators and a mast (if there is not a suitably placed tree available!) it is possible to design and erect practical and effective radiating systems.

Here's wishing you good antenna farming!

John D Heys, G3BDQ

# Chapter 1

# Half-wave dipoles

The origins of the resonant half-wave dipole can be traced back to the successful experiments conducted by Heinrich Hertz in the 1880's. Hertz devised a centre-fed radiator of radio waves, consisting of two rectangular metal plates separated by a spark gap which could be energised either by the discharge of a capacitor (Leyden jars) or by an induction coil. His receiving system was simply a single-turn metal loop which also had a tiny spark gap at its ends. This loop was a tuned circuit which, when arranged to be self-resonant on the same frequency as the 'transmitter', would spark across its gap when near and in the same plane as the two-plate radiator.

Fig 1 represents Hertz's arrangement for his electro-magnetic radiation experiments. The wavelength of his transmissions was largely determined by the length of his transmitting 'antenna' and the area of its capacitive-loading end plates. Using these metal plates would also 'broadband' the antenna, and it made the tuning of the receiving loop much less critical. The physical dimensions of the Hertzian equipment tells us that the wavelengths he used were quite short and lay somewhere in the VHF spectrum.

The later and successful developments in communication by wireless made by Marconi and others centred upon the use of very long wavelengths, so it is not surprising that the concept of the self-resonant Hertz antenna was not examined much through the first 20 years of this century. A half-wave antenna suitable for use on wavelengths of thousands of metres was impracticable to say the least. Hertzian antennas would have been too large even on the wavelengths known at that time as 'short waves' which were allocated to amateur operators in the early 'twenties (around 200m). In Britain there was a 100ft (30m) length restriction on antennas and this inhibited any development of resonant Hertz antennas.

The dramatic amateur DX achievements through and after the winter months of 1923-4, which had only been made possible by the use of shorter wavelengths (between 90 and 100m), were rapidly followed by a further reduction in wavelengths. Much work was begun on the new amateur frequency allocations around 40 and 20m, and then reso-nant half-wave antennas became practical. Amateurs soon abandoned their massive multi-wire Marconi antennas

which were tuned against similar multi-wire counterpoise systems several feet above ground.

The half-wave antennas used by most amateurs in the early 'thirties were often end-fed or centre-fed 'Zepps', using open-wire feedline, or instead were based upon the single-wire feed ideas developed by Bill Everitt and John Byrne at Ohio State University. This latter antenna type became popularly known as the 'Windom' after it was publicised by Loren Windom, the American 8GZ. Centre-feeding a half-wave by using low-impedance feeder had to await the availability of suitable feeder.

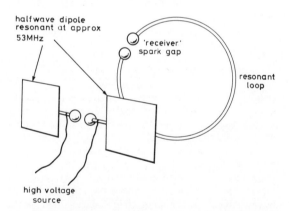

Fig 1. Simplified drawing showing the Hertz transmitter and receiver. With this apparatus Hertz proved that radio waves could be produced and their wavelength determined. He also showed that they could be polarised, refracted and reflected. His 'transmitter' was an end-loaded dipole

Some amateurs used ordinary twisted-pair lighting flex to feed their dipoles but this proved far from satisfactory, for the flex had cotton-covered, rubber-insulated wires which rapidly deteriorated. Even at best when new and dry it had a nominal impedance of 120ohms or more, and was a poor match to a half-wave dipole. It was only during the final years of that decade leading up to the second world war that a practical and effective twin-wire feeder which had a nominal impedance of 75ohms became available. At the same time the low-impedance screened cable which we now call 'co-ax' was developed for the infant television industry.

The introduction of low-impedance feeders enabled that universal workhorse, the half-wave dipole, to also make its debut. Today the half-wave dipole antenna has become the standard against which other radiating systems are judged and it remains as perhaps the most effective yet simple single-band antenna, and one which can be guaranteed to perform well even when used in far-from-ideal situations.

## Dipole basics

A resonant half-wavelength of wire will be somewhat shorter than its name implies. RF energy in free space (electromagnetic radiation) can travel at the speed of light, but when moving along a conductor it goes more slowly. At HF (between 1.8 and 30MHz) wires exhibit 'skin effect' – most of the RF energy flows along the outer surface of the conductor. A practical half-wave antenna made from wire needs end supports; each end usually being terminated at an antenna insulator. The capacitance between the ends of dipole and its supports, even when the supporting material is non-metallic, gives rise to 'end effect'. This effect additionally loads the wire capacitively and contributes towards its shortening from the theoretical half-wavelength.

The theoretical half-wavelength may be calculated from the expression:

$$150/f(\text{MHz}) \text{ metres} \quad \text{or} \quad 492/f(\text{MHz}) \text{ feet}$$

These statements are useful in calculating the spacing between elements in beams and so on, but for most practical purposes on the HF bands a modified formula which takes into account end effects etc must be used when determining the lengths of actual half-wave antennas. The length of a typical half-wave antenna is given by $468/f(\text{MHz})$ feet. Any length in feet may be easily converted into metres by multiplying that length by 0.3048; or instead a direct calculation of a half-wavelength in metres can be made by using $143/f$ (MHz) metres.

L A Moxon, G6XN, favours the suspension of wire antennas with nylon ropes, a practice which obviates any need for end insulators. In his book *HF Antennas for All Locations* (published by the RSGB), Moxon suggests that when no insulators are used a half-wavelength can be found by using either $478/f(\text{MHz})$ feet or instead $145.7/f(\text{MHz})$ metres.

A further factor which influences antenna resonant length is the diameter of the wire used for that antenna. The formulae already given are accurate for wire antennas and it is only when tubing or extremely thin wire (which would be impractical anyway) is used that different calculations are needed. The author is confining his attentions to antennas made from wire, and their lengths are shown in

Table 1. This provides length details for the amateur HF bands both when using insulators or nylon rope.

**Table 1. Lengths of half-wave dipoles**

| Frequency | Length | | | |
|---|---|---|---|---|
| | With insulators | | Without insulators | |
| (kHz) | (ft) | (m) | (ft) | (m) |
| 1850 | 252' 11" | 77.29 | 258' 5" | 78.75 |
| 1950 | 240' 0" | 73.33 | 245' 1" | 74.71 |
| 3550 | 131' 10" | 40.28 | 134' 8" | 41.04 |
| 3750 | 124' 9" | 38.13 | 127' 5" | 38.85 |
| 7050 | 66' 4" | 20.28 | 67' 10" | 20.66 |
| 10,100 | 46' 4" | 14.15 | 47' 4" | 14.42 |
| 14,100 | 33' 2" | 10.14 | 33' 11" | 10.33 |
| 14,250 | 32' 10" | 10.03 | 33' 6" | 10.22 |
| 18,100 | 25' 10" | 7.90 | 26' 5" | 8.04 |
| 21,100 | 22' 2" | 6.77 | 22' 8" | 6.90 |
| 21,300 | 21' 11" | 6.71 | 22' 5" | 6.84 |
| 24,940 | 18' 9" | 5.73 | 19' 2" | 5.84 |
| 28,100 | 16' 8" | 5.08 | 17' 0" | 5.18 |
| 28,500 | 16' 5" | 5.01 | 16' 9" | 5.11 |
| 29,000 | 16' 1" | 4.93 | 16' 6" | 5.02 |
| 29,500 | 15' 10" | 4.84 | 16' 2" | 4.93 |

## Impedance

Newcomers to amateur radio often find it confusing to distinguish between the terms 'resistance' and 'impedance'. As impedance is so important when discussing antennas a few words on this topic will, it is hoped, not be out of place. Resistance in units of ohms can only be related to non-inductive or non-capacitive devices, ie materials such as metals (wires) and carbon etc which restrict the flow of current in an electrical circuit. These materials may be fabricated into the many varieties of resistor available and will exhibit the same restrictive properties to either direct or alternating current. At very high AC frequencies, however, other factors such as the skin effect may complicate matters, but in general ordinary 'resistances' are equally effective when the currents are either AC or DC.

Any inductance or capacitance will restrict the flow of AC current. The restriction, termed 'reactance', will depend on the value of the inductance or capacitance and the frequency. The higher the frequency, the greater will be the reactance of the inductor and the lower the reactance of the capacitor. The combination of reactance and resistance is termed 'impedance'. Pure resistors have no reactive components and hence their impedance is equal to their resistance.

In a circuit comprising both inductive and capacitive reactances, one tends to counteract the other – if the reactances are equal a condition termed 'resonance' exists. If, for an impedance comprising resistive, capacitive and inductive elements, the capacitive and inductive elements are in resonance then the impedance will become purely resistive. This forms the basis of tuning antennas to resonance. Impedance is measured in ohms and a more detailed explanation of these matters may be found in most of the handbooks on amateur radio.

## Dipole impedances

A half-wave transmitting antenna, when energised and resonant, will have high RF voltages at its ends with theoretically zero RF currents there. This means that the ends of a half-wave dipole in free space will have an infinitely high impedance, but in practice down here in the real world there will always be some leakage from its ends and into the supporting insulators etc. This means that in reality the impedance at the dipole ends is close to 100,000ohms, a value which depends upon the wire or element thickness. At a distance of approximately one-sixteenth wavelength from either end it is 1000ohms, and at the dipole centre, where the current is greatest and the RF voltage is low, the impedance is also low.

If it were made from an infinitely thin conductor wire, our theoretical dipole in free space would have at its centre an impedance of about 73ohms. Such an antenna is impossible in the material world, and a practical half-wave dipole made from wire will have an impedance at its centre at resonance close to 65ohms. Antennas fabricated from tubing have lower values at their centres, of between 55 and 60ohms. These impedance values depend upon the height of the antenna above ground (see later).

The very high values of self-impedance at the ends of a half-wave wire makes end-feeding difficult, and this is why breaking the wire at its centre and connecting the inner ends so formed to a low-impedance feedline makes a convenient and efficient coupling and match. Suitable feeder is available in the form of twin-lead or coaxial cable, which both have design impedances lying between 50 and 75ohms. These present a good match to dipole centres.

At exact resonance the impedance at the centre of a half-wave dipole is like a pure resistance. At any other frequencies, however, the same dipole will have either inductive or capacitive reactance at its feedpoint. If the dipole is too short to be resonant the reactance is capacitive and when it is too long the reactance becomes inductive. In either case there will be problems in matching the 70ohm feeder to the dipole and, if the reactances are great, there will be a high SWR on the feeder and considerable power loss.

## Antenna Q

At resonance there will appear to be pure resistance at the dipole centre feedpoint but, if the antenna is too long or too short for the frequency in use, some reactance will also be present there. Half-wave dipoles which are too long will exhibit inductive reactance at their feedpoint, and those too short will show a capacitive reactance. This reactance will make it more difficult to feed the wire or effect a full transfer of power from the feeder. The mismatch will also result in SWR readings above unity, and these will become progressively worse as the antenna becomes more off-tune.

A half-wave antenna is something like a conventional tuned circuit where the $Q$ or 'magnification factor' is largely determined by the resistance of the coil. Losses in the capacitor used in the circuit are generally small and are not so significant in the determination of $Q$. A high-$Q$ tuned circuit exhibits very sharp tuning (selectivity) and this is also the case when an antenna has a high $Q$.

Using thin wires lowers the bandwidth of a half-wave antenna but not dramatically. However, short wires that are brought into resonance will exhibit high $Q$. The shorter the wire in terms of wavelength, the higher the $Q$. Small changes in the transmitting frequency away from the antenna resonances will give rise to a rapid rise in the reactance at the feedpoint.

Thicker wire will lower the $Q$, reduce ohmic loss and make the half-wave dipole less frequency conscious. It is therefore best to ensure that such an antenna is made from the thickest possible wire consistent with such factors as the pull on the antenna supports, windage and sag.

## Dipole height

The height of a horizontal dipole above the ground as a ratio of its design frequency is important (see the standard curves of feed impedance against height in Fig 2). When below about half a wavelength high the radiation resistance at the feedpoint will be reduced, and down at a height of just one-tenth of a wavelength it will only be 25ohms. This means that a dipole fed from a standard type of low-impedance feeder will suffer a considerable mis-match when near the ground. One-tenth of a wavelength is about 50ft (15m) on 1.8MHz and as little as 3 to 4ft on 28MHz. This helps to explain why low dipoles on the lower-frequency bands are far-from-efficient radiators.

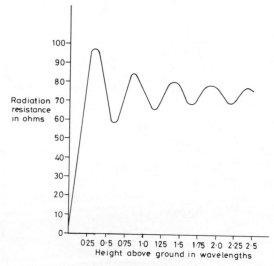

Fig 2. Radiation resistance of a half-wave dipole as a function of height above the ground *(The ARRL Antenna Book)*

A horizontal half-wave antenna, if at least a half-wavelength above ground, will radiate most of its applied power at right angles to the line or axis of the wire. Its radiation pattern may be visualised as having the shape of a torus or doughnut, with the wire running through the centre hole (Fig 3). About 40° on either side of the broadside maxima of radiation the power falls to half (3dB down), and it will fall rapidly as the angle increases.

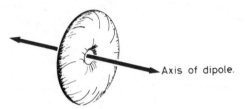

Fig 3. A crude representation of the radiation from a half-wave dipole antenna in free space. Practical antenna systems are, however, influenced by ground reflection and their radiation patterns are much modified

Theoretically there should be little or no radiation off the wire ends but in practice there will remain some radiation at high angles to the horizon from both ends which might prove useful for short-range work. The horizontal radiation pattern at both 30° and the low angle of 9° may be seen in Fig 4. The 30° high-angle radiation from a half-wave dipole at a height of half a wavelength will tend to be from one to two S-points greater than the low-angle radiation needed for DX working in most directions (that

is about 5 to 10dB better), and it emphasises the fact that a half-wave dipole is a general-purpose, 'all-round' antenna type, good for both semi-local and distant working.

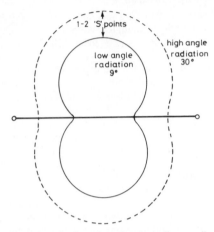

Fig 4. The horizontal polar diagram of a half-wave dipole at a height of half a wavelength above the ground, showing the considerable high-angle radiation (at 30°) off the ends of the wire. The low-angle radiation is mainly at right angles to the wire and is from 5 to 10dB down relative to the 30° radiation

From the ends, however, there is little low-angle radiation, and here it is as much as three to four S-points down from the maxima at right angles to the wire. This explains why a dipole is best arranged to be at right angles to the

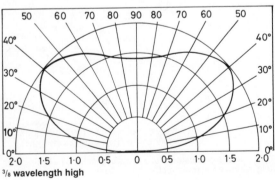

1/8 wavelength high

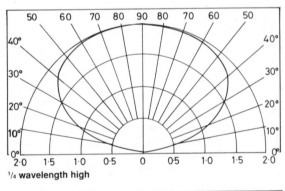

1/4 wavelength high

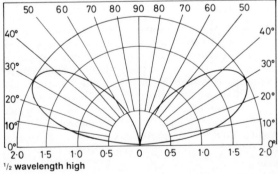

3/8 wavelength high

1/2 wavelength high

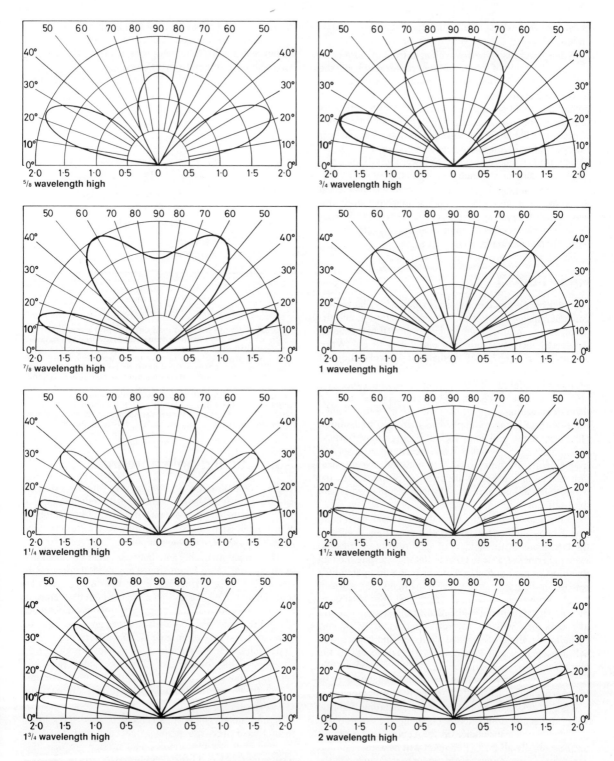

Fig 5. Vertical radiation patterns of horizontal antennas at differing heights above a perfectly conducting ground *(The ARRL Antenna Book)*

areas to be worked, especially for long-distance communication.

When lower than a half-wavelength from the ground, a greater proportion of the transmitted power will leave the antenna at high angles. Such a dipole at a quarter of a wavelength above ground or lower will have almost no DX potential but will radiate 'sky waves'. For some amateurs this can be a boon for there are many who only wish to place strong and reliable signals into all parts of the UK and near Europe. Users of the many 'natter nets' on the 3.5 and 7MHz bands would be advised to lower their dipoles to a height of a quarter-wavelength or under!

The vertical radiation patterns of horizontal antennas at different heights above the ground ranging from an eighth-wavelength to two wavelengths can be seen in Fig 5. In all these examples it is assumed that the antennas are above a perfectly conducting ground.

# Radiation

A novice once asked the author to explain why a half-wave antenna radiated most of the power applied to it, and the prompt reply informed him that after the RF had gone up the feeder there was nowhere else for it to go! When attempting a simple non-mathematical explanation of antenna radiation it is perhaps in order to first think of an ordinary parallel-tuned circuit (Fig 6(a)), consisting of a coil and a tuning capacitor. When not enclosed in some form of screening compartment this circuit will radiate a little of the RF energy applied to it from a transmitter. However, if the coil is of the large single-turn variety, much more energy is radiated and the circuit behaves as a loop antenna. In the case of a normal coil, the physical dimensions of the circuit are small in proportion to the wavelength and much of the RF field is contained within the oscillatory circuit.

In the years prior to the second world war, when there was the so-called 'artificial aerial' transmitting licence in the UK, many of the holders of such restrictive 'tickets' were nevertheless able to contact stations, despite the fact that most of their output power was being dissipated by a resistor in the tuned circuit which made up the 'artificial aerial'. With such a circuit, if unscreened, it is possible to receive signals when it is coupled to the station receiver although they will be severely attenuated. All wires, including straight ones, have both self-inductance and self-capacitance, and if two wires are connected to the tuned circuit (Fig 6(b)) that must then be reduced in inductance and capacitance if the same frequency is to be tuned. The radiation from the circuit will now be enhanced.

This process may be continued by extending both the wires further and simultaneously reducing the values of the inductance and capacitance in the central tuned circuit until eventually all that will remain will be a tuned circuit consisting of two wires which couple to the power source via a small, single-turn link coil (Fig 6(c)). Lo and behold!

We now have a half-wave dipole. This simple explanation of antenna resonance and radiation is perhaps not very elegant but it helps to explain what happens when a half-wave dipole is driven from an RF source.

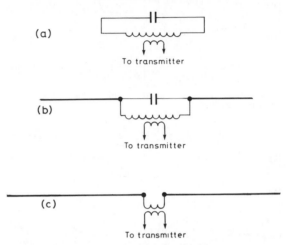

Fig 6. (a) A simple parallel-tuned circuit using a multi-turn coil which, when coupled to the output of a transmitter, will radiate a small proportion of the circulating power. The larger the tuned circuit is in physical terms relative to the wavelength used, the more RF will be radiated. (b) If short equal lengths of wire are connected to each end of the tuned circuit, their self-inductance and capacitance will allow a reduction of the tuned-circuit L and C constants, and a greater proportion of the applied power will be radiated. (c) If the wires are lengthened to such a degree that the centre tuned circuit is no longer necessary, most of the applied power will be radiated and we have a half-wave dipole antenna

# A practical dipole antenna

There is no doubt that the simplest and yet most effective all-purpose, single-band antenna for the amateur, which can be guaranteed to work well without trimming or tuning adjustments so long as it conforms to the basic design parameters, is the half-wave dipole.

The top length of this antenna may either be calculated or taken from Table 1, and almost any kind of copper wire can be used. For permanent or semi-permanent installations 16 or 18SWG hard-drawn copper is to be preferred, and for experimental and temporary antennas most types of stranded and plastic-covered wire can be used. The resonant frequency of an antenna made from this wire is said to fall by 3 to 5% but the author has never noticed such an effect. This lowering of frequency may, however, become noticeable when making and testing wire beams, especially those with quad loops as driven and parasitic elements.

If any antenna end insulators are to be used, those made from Pyrex glass are perhaps the best, and when the antenna ends are close to a metal mast more than one insulator at each end is to be preferred. Nylon or Terylene cords may be tied directly to the antenna wire ends

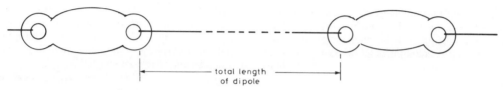

Fig 7. The end insulators of a dipole antenna, showing that the total length of the antenna includes the wrap-round at the tie points. Although small, such additions could detune the antenna on the 21 or 28MHz bands

(knotted), and they will make effective insulators if more than about 2m in length.

The antenna maker must bear in mind the change in resonant lengths which are induced by this technique (see Table 1), and cut his dipole longer than it would be if normal end insulators were to be used. Without end insulators the suggested 2m or more of cord between the antenna and the fixing point will present a very low leakage path even in wet weather. Although ultra-violet energy (present in sunlight) can bring on a deterioration in the structure of non-organic ropes, the author has had nylon cords in use for many years with no apparent ill effects.

If antenna insulators are used the length of the resonant top of the antenna must also include the furthest ends of the loops which pass through the insulators (Fig 7). The inch or so involved here can be important on the higher-frequency bands, but of course is nothing to worry about on 3.5MHz! If joins are required in the top wire it is best to solder and tape or weatherproof them. Try to avoid joins towards the centre of the dipole, for here the RF current is high and any ohmic losses will reduce radiated power. Experience has taught the author that twisted joins, not soldered, are satisfactory when made at a high-impedance point such as towards the far end of an antenna, but the purists will no doubt disagree!

The dipole top is broken at the half-way point and here an insulator must be inserted. This point is at low RF potential and low impedance so the insulation need not be high. Using expensive glazed ceramic or glass centre insulators is a waste of money and most plastics such as Perspex, acetate or similar insulating material may be used. The centre blocks are best fabricated in the shape of a 'T' (Fig 8) or a 'Y' so that there will be some way to securely anchor the top few inches of feeder. When using either twin-wire 72ohm feeder or the heavier coaxial cable there must not be any strain put on to the connections to the dipole wires. All antennas will sway or swing even during relatively calm weather conditions; this can easily induce metal fatigue and an early demise of feeders high up where they cannot be observed. Fig 8 shows suggested anchoring methods for both types of feeder.

The dipole feeder must run down vertically for at least a quarter-wavelength before it bends to run to the house or shack, and if possible it should avoid running below and in line with the antenna top. A feeder beneath the leg of a dipole antenna will unbalance the system and will lower feedpoint impedance. A useful and tidy way to arrange a coaxial feeder is to run it vertically down from the antenna to the ground and then bury it a few inches down on its run to the shack.

Unscreened twin-wire feeder must not be buried or its nominal impedance may be affected, and there will then be a high SWR on the line. The twin feeder is, however, less susceptible to dampness effects than some of the 300ohm

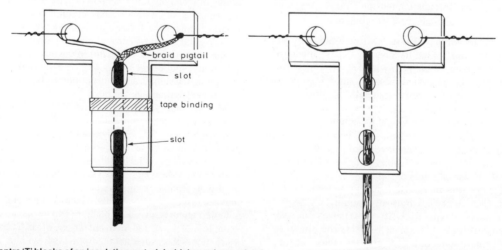

Fig 8. Centre 'T' blocks of an insulating material which may be used at the centre of a dipole antenna. Coaxial-cable feeder, being heavier, will require more support than the twin-wire 72ohm type of feeder. Almost any insulating material which is weatherproofed may be employed for the centre blocks. No strain must be put on the soldered connections to the dipole halves.

ribbon cables available, for most of the electrical field between the conductors is confined within the solid black polyethylene insulating material into which they are embedded. The black colouring helps to reduce UV damage (caused by sunlight) to the plastic, so avoid light-coloured or transparent cable varieties.

When coaxial cable is used to feed the dipole some extra care is needed where it connects to the antenna. The outer jacket of the cable should be stripped for about 4in. and then just above the new termination of the outer sleeve a hole is made through the copper braid. The insulated 'inner' may then be pulled out through this hole. In this way the braid 'pigtail' will be strong and unlikely to fray away or weaken, especially if its end is tinned with solder. The feeders must be soldered to the antenna wires and then the joints taped or similarly weatherproofed.

There are some amateurs who deprecate coaxial feed to dipoles and advise the use of 1:1 baluns at the feedpoint. Although coaxial feed does not result in a true balanced system, it actually does work well in practice without resort to baluns. It is, however, especially important that any coaxial feeder comes down vertically from the feedpoint, and then preferably lies along or under the ground on the remainder of its run.

Some arguments that are put up against using unbalanced coaxial feeders include the following statements:

1. There may be RF currents induced onto the outer braid of the cable.
2. The outer shield may radiate.
3. There is a greater sensitivity to nearby objects such as masts, telephone and other overhead wires, buildings or trees.
4. Losses can be high if the cable weathers badly and inside corrosion begins.
5. Coaxial cable is heavy and can pull the top wire down considerably, so inducing tension strains.
6. There may sometimes be TVI problems.

A twin-lead feeder may also suffer from weathering and a better type to look for has its inner conductors made from enamelled copper. With twin-wire feeder, which has a nominal impedance around 72ohms, an ATU becomes almost obligatory for correct matching to most modern transceivers. TVI problems are, however, fewer using this kind of feeder and it is very light in weight.

For some reason 72ohm twin-wire feeder is becoming difficult to locate, and when advertised it costs almost as much per metre as coaxial cable. Line losses with either twin-wire feeder or good coaxial cable are negligible for average cable runs on the HF bands up to frequencies of 30MHz. Coaxial cable will usually (not in the case of the expensive special types for UHF work) have a line loss of less than 2dB for a 100ft run at 10MHz. The loss in twin lead is similar to that of a good coaxial cable.

Waterproofing the cable, especially when using coaxial cable, is very important. Some people use the so-called

'VHF' plugs and sockets (PL-259 and SO-239) to make connections to the centre of dipole antennas. This practice can invite disaster, for such connectors are far from being weatherproof. They also add to the cable weight and help to drag the antenna down into a 'V'.

After soldering the coaxial or twin feeder to the antenna the cable ends should be thickly coated with a strong and adhesive waterproofing material. Probably the best substance for this is the silicone-rubber sealant now available from most hardware stores. The author prefers the clear, colourless form and always takes care to ensure that everything is well covered and that there are no small remaining pin holes caused by air bubbles being trapped in the sealant.

Where there is no restriction in the siting of the antenna, and the use of only one dipole is contemplated, it is best to run it from north to south. In this way most of the world will be covered, there being but few densely populated areas lying due north or south of the UK. A pair of half-wave dipoles arranged to be at right angles to each other would be ideal, and if possible an arrangement with one running NW/SE and the other from NE/SW should be tried. An examination of a great circle map shows that such a disposition is very good for full world coverage when using half-wave dipoles.

An extra small dividend is available when a dipole is cut for the 7MHz band for then it will also work fairly well as a centre-fed $1^1/_2$-wave wire on the 21MHz band. It will, however, have quite a different radiation pattern to a standard dipole, but show a little gain over a dipole cut for 21MHz in its preferred directions of radiation.

## Matching

Mention was made earlier of the use of an ATU (antenna tuning or matching unit) when twin-wire feeder is used. In the author's opinion an ATU should always be used whatever antenna or feeder type is available. An ATU can only do good, and it will provide an additional tuned circuit on the transmitting or receiving frequency which will aid selectivity, reduce cross-modulation effects from strong off-frequency signals and, most importantly, cut down any radiation of harmonics or spurii.

Modern solidstate transceivers are likely to suffer damage if worked into load impedances not far removed from a nominal 50ohms. Some designs will automatically reduce power output when working into a mismatch in the order of 3:1 or even less. The use of an ATU can help to prevent such problems arising, especially when a dipole is being used away from its design frequency and therefore does not match correctly to its feeder. Such a mismatch to the antenna cannot be avoided and there will be of course some small loss of power radiated, but the use of an ATU in such a situation will ensure that the transceiver will be 'fooled' into behaving as if all is unchanged despite the mismatch.

An ATU should of course also be used in conjunction

with an SWR meter, and if this instrument is connected between the end of the feeder and the ATU, any mismatch which occurs when the dipole is used on either side of its design frequency will be very noticeable. At best, even on the dipole's resonant frequency, it is unlikely that the SWR reading will be perfect unity, for it is very difficult to achieve a perfect match to either 50ohm or 72ohm at the dipole centre. This is because the nominal antenna impedance is so dependent upon such factors as height, the type of ground beneath the antenna, nearby objects and so on.

Many amateurs, especially when quite new to transmitting, worry themselves unduly about the antenna-line SWR readings. It must be remembered that older hands operated very successfully for many years through times when SWR meters did not even exist! It needs an SWR reading of 3.7:1 to double feedline losses and it is unlikely that even the most ill-fashioned or awkwardly positioned dipole would present such a high mismatch. If the SWR reading lies between unity and 1.5:1 any mismatch loss will be negligible. At such an SWR reading the total line losses will then only be the cable losses multiplied by a factor of 1.1!

Before leaving the topic of SWR on feedlines it must also be stressed that a feeder connecting to a half-wave dipole antenna can be of *any* length, so long as its nominal impedance equals the impedance at the dipole centre. If any addition or subtraction of feedline greatly affects the SWR present, it means that there must be a serious mismatch at the antenna connection. Some of the more simplistic articles which were once prolific in CB magazines actually advocated the technique of trimming coaxial cable length in order to bring the SWR down to unity!

## Broadbanding the dipole

It has already been stated that a low-$Q$ half-wave wire will have a wider bandwidth than a high-$Q$ one which has been made with very thin conductor wire. One way to ensure that a dipole covers an entire band (especially on the 3.5MHz

and 1.8MHz bands) is to use a very thick wire for the antenna. Two or three thinner wires can be put in parallel to achieve this but a better way is to use a 'fantail' or 'bow-tie' arrangement (see Fig 9), which will provide a low SWR right across an LF band.

On the higher frequencies any normal antenna wire will of course be thicker in terms of wavelength, and therefore antenna $Q$ will be lower. Despite this, those who intend CW operation at the LF end of the 28MHz band, together with some working on FM or satellite reception above 29MHz, might find broadbanding an advantage. A half-wave wire cut for 29MHz is almost a foot shorter than one for 28MHz; a difference of about 6%.

The outer wires in a three-wire 'fantail' arrangement are made a little shorter than the centre wire and they will resonate at the HF end of the band for which they are cut, whereas the centre wire can be cut to resonate at the LF end. The wire arrangement can be altered of course, so that the shorter wire lies in the centre between the two LF wires, or instead all three wires can be of different lengths, so covering HF, LF and mid-band frequencies. A separation of 3 to 4ft between the wires is adequate, and they may either go to different end tie points or more conveniently attach to plastic or wooden spreaders (via insulators). Table 1 can be used for determining the wire lengths of broadband multi-dipoles.

## Sloping dipoles

Horizontal half-wave dipoles require two end supports and it is not always possible to provide these in some awkward locations. In such situations a single support, preferably a non-metallic mast or a high point on a building, will suffice, and then the antenna can be arranged to slope down towards the ground at an angle lying somewhere between 45 and 60° (Fig 10). The sloping half-wave dipole must have its lower end at least one-sixth of a wavelength above ground, and its feeder must come away from the radiator at 90° for at least a quarter of a wavelength too. If coaxial

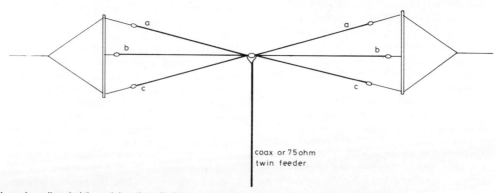

coax or 75ohm
twin feeder

Fig 9. How a broadbanded 'bow-tie' or 'fantail' dipole can be constructed. The three dipoles a, b and c are cut to resonate at the band edges and centre band frequency. This technique will allow a single antenna to be used on the 3.5MHz band. A half-wave dipole cut to resonate at the LF end of this band will be 7ft (2.13m) too long at the HF end. On the higher-frequency bands the width of the band relative to the frequency is small, and a single dipole cut to mid-band will suffice

feeder is used the metal braid must connect to the lower half of the antenna to help feed balance.

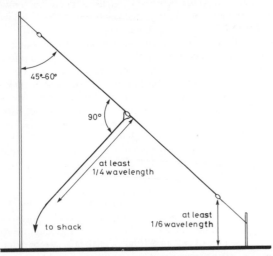

Fig 10. A half-wave sloping dipole which can be put up in a small space and which will be useful for long-distance working. Most of its low-angle radiation is towards the low end of the antenna but there is also a considerable radiation at high angles in other directions. The feeder must come away from the antenna at right angles for at least a quarter-wavelength. If possible non-metallic masts should be used to support a 'sloper', but when this cannot be arranged ensure that the mast length is not close to a half-wavelength at the operating frequency

The performance of a sloping dipole is quite different from one of the horizontal variety and it can be good for DX work. The radiation from a sloping dipole shows slant polarisation with both vertical and horizontal components according to the amount of slope. Its lower angle of radiation to the horizon can result in useful DX gain over a horizontal dipole. Some workers claim this gain may lie between 3 and 6dB but others give lower figures.

This compares favourably with some of the cheaper multi-element 'trapped' beams which often have a performance on their lowest frequency band inferior to that of a dipole! This kind of gain is difficult to realise on the LF bands in other ways, and for most amateurs multi-element Yagi beams are out of the question.

There is some high-angled radiation from the sides of the sloping dipole but very little radiation from its high end. An actual plan of the horizontal radiation pattern resembles a heart with a null between its two upper lobes. This null corresponds with the high end of the sloping dipole. A disadvantage is of course that long-distance working will only be possible towards one direction, but this may be overcome by having a group of three or four 'slopers' suspended from a common central support, each with its individual feedline which may be selectively switched to the transmitter or receiver.

There are designs which involve the unused dipoles in such arrangements as reflectors to improve forward gain

and front/back ratios, but their construction and adjustment can be complicated and they lie outside the scope of these pages.

## Vertical dipoles

A vertical half-wave dipole will radiate vertically polarised signals all round, and much of the radiation will be at the low angles favourable for DX working. Unfortunately the centre of this antenna must be at least 0.45 wavelength above the ground if a feed impedance of around 70ohms is needed. Instead it is usually more convenient to arrange for a vertical quarter-wave antenna to be used which can then have its feedpoint at or near ground level.

Vertical half-waves are not often used by amateurs, although they become practical on the 28MHz band and then can be suspended from existing wires or support cords. A vertical dipole must have its feeder coming away from the radiator wire at right angles, and this may present some problems. They cannot be hung down from metal masts or towers either, so their applications are rather restricted. Experiments by the author using suspended vertical dipoles on 21 and 28MHz showed that vertical quarter-wave radiators using ground planes were more effective, so half-wave versions cannot really be recommended.

## Monoband and multiband inverted-V dipoles

The maximum radiation from any antenna is from the points of high RF current, and a half-wave dipole has this maximum at its centre and for a few feet on either side of the feeder connections. A horizontal dipole made with wire will invariably have some centre sag and, if coaxial feeder is used, this can be pronounced and will lower the effective height of the antenna. High end supports do not help in this situation. However, if the centre of the dipole can be elevated and both dipole halves are made to come down in the form of an inverted-V, the point of maximum current and therefore the maximum radiation will be at the highest point above the many surrounding objects which might screen the antenna.

It is often possible to have one fairly high mast in the centre of a garden or plot in locations where the erection of a pair of similar supports with their attendant guying wires would be difficult. If the operating position is located close to such a central mast it will then be ideally placed to receive the feeder from an inverted-V. A roof-mounted or chimney-mounted mast may also serve as the centre support for a 'V', and the two ends of the dipole can then drop down on either side of a house or bungalow roof. Such chimney mounting will allow the feeder to be dropped to the shack quite easily when the operating position is in the house.

An inverted-V antenna has its greatest radiation at right angles to the run or axis of the sloping wires, but experience has taught the author that this antenna does not have such pronounced nulls at its ends as a conventional horizontal dipole. The side radiation is at quite low angles and is good for long-distance working.

The inverted-V has indeed an excellent reputation for DX communication on the lower-frequency amateur bands where the erection of large verticals or high horizontal dipoles is not practicable. There are, however, some design features concerning this antenna which must be considered when contemplating making one.

The angle between the sloping wires must be *at least* 90° and preferably be 120° or more. This angle dictates the centre support height as well as the length of ground needed to accommodate the inverted-V. When designed for the 3.5MHz band an inverted-V will need a centre support at least 45ft (14m) high and a garden length of around 110ft (34m). A horizontal dipole needs at least 132ft (40m) of garden and that neglects to take into account guys to the rear of the end support masts. Portable work on the 14MHz band using a thin-wire inverted-V will only need a lightweight 15ft (5m) pole to hold up its centre!

The sloping of the dipole wires causes a reduction of the resonant frequency for a given dipole length so about 5% must be subtracted from standard dipole dimensions. One reason for this is the increased self-capacitance of the antenna when its ends are brought closer together and also towards the ground. The self-inductance of the wire is also increased, for the inverted-V approximates to almost a half-turn loop. The calculated wire lengths for inverted-V dipoles on the amateur bands are given in Table 2.

Yet a further consequence arising from sloping the dipole wires is a change in its radiation resistance. The centre impedance falls from the nominal 65ohms of a horizontal dipole to just 50ohms. This of course is ideal for matching the antenna to standard 50ohm impedance coaxial cable! An inverted-V antenna has a higher *Q* than a simple dipole so it tends to have a narrower bandwidth.

It is not recommended that the ends of an inverted-V come closer to the ground than about 10ft, even on the higher-frequency bands, for there can be a possible danger

to a child, other unsuspecting person or animal touching its ends which will be at a high RF potential when energised. The effects, although not likely to prove lethal, nevertheless could result in a nasty shock or RF burn, and it seems unlikely that an insurance company would look kindly at any claims resulting from such an accident.

**Table 2. Lengths of inverted-V dipoles**

| Frequency (kHz) | Length | |
|---|---|---|
| | (ft) | (m) |
| 3600 | 123' 6" | 37.74 |
| 7050 | 63' 1" | 19.27 |
| 10,100 | 44' 0" | 13.45 |
| 14,200 | 31' 4" | 9.57 |
| 18,100 | 24' 7" | 7.69 |
| 21,200 | 20' 11" | 6.41 |
| 24,940 | 17' 10" | 5.45 |
| 28,200 | 15' 9" | 4.82 |
| 29,200 | 15' 3" | 4.65 |

Coaxial feed is recommended with an inverted-V, and the low-loss heavier varieties of cable can be used to advantage, for there are no sag problems when the feeder is fastened up at the top and also down the length of the mast. The feeder will impose no strain upon the antenna or the soldered connections at its feedpoint. A light 25ft (8m) mast made up from bamboo lengths was once used by the author to hold up a 7MHz inverted-V antenna, and even this structure was strong enough to carry TV-type coaxial cable without difficulty. No balun was found to be necessary when using any of the several antennas of this type put up by the author. Twin-wire 72ohm feeder may be used instead of coaxial cable, and this will certainly call for an ATU at the transceiver end of the feeder. The general arrangement of an inverted-V dipole is shown in Fig 11.

## Multiband dipoles

All the dipoles discussed so far in this chapter have been

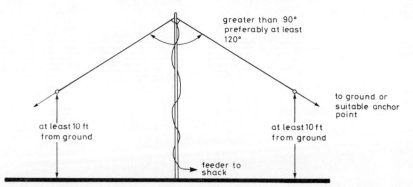

Fig 11. A half-wave inverted-V dipole with the angle between the top wires at 120°. This angle must never fall below 90°. The centre support mast puts the high RF current section of the antenna at the highest point and also carries the weight of the antenna and the feeder. The inverted-V is good for DX working and will give good results on the 3.5MHz band when the mast is only about 45ft (14m) high

monoband devices, but the inverted-V configuration offers a choice of either monoband or multiband working. For example, if two half-wave dipoles are cut to resonance on different wavebands and are then centre-fed by a common feeder, each will present a feed impedance of around 65ohms at its own particular resonant frequency and will not be affected by its neighbour.

A dipole cut for 7MHz will have a normal and low centre impedance on that frequency but, should a second dipole which has been cut for 14MHz also be connected to a common feedpoint, this second and shorter dipole will not present a centre impedance able to accept power at the lower frequency of 7MHz. The converse will apply when the 14MHz dipole is driven, for then the 7MHz dipole will become a centre-fed full-wave antenna and its high centre impedance will not affect the working of the shorter 14MHz wire.

In this way several dipoles cut for different frequency bands may be connected in parallel with each other, so long as they are even multiples of the antenna which operates on the lowest frequency. The shorter dipoles will all present high impedance at the centre feedpoint when they are not being driven.

If one of the dipoles is resonant on 7MHz a 21MHz dipole (3 x 7MHz) need not be included in the parallel arrangement, for the 7MHz antenna will also radiate on the higher band as a centre-fed 1½-wave wire.

A multiband antenna using a number of horizontal dipoles can be devised in the way suggested, but the extra weight over that of a single dipole will cause considerable sag and a fall in the effective height of all the dipoles. However, it is possible to make an effective multiband system using the inverted-V configuration with its central high point where the feeder and the combined weight of the dipoles may be anchored.

Should more than two or three dipoles be used as a multiband inverted-V, their wire ends must come down to separate anchor points and this makes a rather untidy arrangement. A more elegant solution which will allow the construction of a lightweight, five-band inverted-V, and which has single tie points at each end and a common angle at the antenna top, involves the use of flat multi-way ribbon cable for the radiators.

One easily available type of this cable is 10-way, and each of its conductors is made from 14 x 0.13mm tinned copper strands. This cable is 13mm wide and only 1.3mm thick. Each conductor wire is rated at 1.4A continuous current, which means that if used as a half-wave antenna it could handle almost 140W of RF power. The short duty cycles of CW (50%) and SSB operation would enable such a conductor wire to cope quite easily with full UK legal power.

The cable can be obtained in complete lengths of up to 50m, and a five-band inverted-V using about 40m could be made up from one such piece. This antenna would operate on 3.5, 7, 14, 21 and 28MHz. Using Table 2 to find the individual dipole lengths, parts of the cable can be cut away so that the ribbon becomes progressively narrower and lighter towards its ends; each section left being a quarter-wave for the separate bands (leaving out a dipole for 21MHz for the reasons already given). As the ribbon has 10 conductor wires its adjacent pairs can be paralleled at each end, so doubling the power handling capability and in addition broadening the bandwidth by lowering the Q of each dipole. The ribbon inverted-V is shown in Fig 12 (not to scale).

If this suggested form of construction is used for the antenna the usual method of end-tying to insulators or a nylon cord will not suffice, for then too much strain will be put on to the longest dipole in the group. It will have to carry its own and also the weight of the other three dipoles, and is likely to break when stressed. A method used successfully by the author is to buy a 50m length of thin (1mm diameter) but strong nylon cord, put one end of this through the eye of a stout packing needle and 'stitch' it along the multi-way ribbon cable. The stitches can be quite

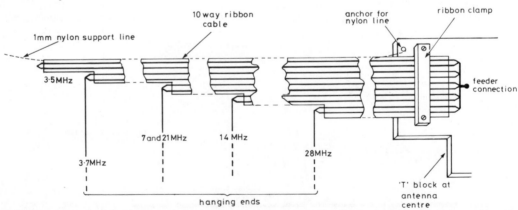

Fig 12. A simplified diagram showing how a 10-way cable may be used to construct an effective five-band inverted-V antenna. The thin nylon line used to support the ribbon is 'stitched' along the length of the antenna. By letting the ends of the dipole sections hang down for a few feet, capacitive and other interaction between them is minimised. There is no need to drop the ends of the 3.5MHz dipole. This drawing is not to scale

large, each one being about 2ft long, and the end product will be a strong and lightweight multi-dipole top which can be easily supported by a normal 40 to 45ft (12 to 14m) pole or mast. The ends can be secured by the thin nylon cord so that the top angle is the recommended 120° or so.

The ends of half-wave dipoles are at quite high RF voltages, and their proximity to other wires etc will capacitively detune them. It is therefore expedient to allow a few feet of the 7, 14 and 28MHz dipole ends to dangle freely away from the remaining ribbon. A good centre connecting block of Perspex or similar material will make a suitable anchor point for the ribbon ends and also the feeder (Fig 12). The ends of the longest dipole sections must not come right down to ground level but should be arranged to terminate at about 10ft (3m) and above the ground for reasons discussed earlier concerning the simple inverted-V antenna.

A multiband inverted-V made on these lines gave very good results at the author's QTH, and without any 'trimming' or other adjustment SWR readings between 1.3:1 and 1.7:1 were achieved. The worst reading was on 21MHz when working into the 7MHz dipole section – this behaved as a 1½-wave wire and had then a feed impedance a little higher than 50ohms.

Simpler versions of the antenna for use on just two or three bands may be constructed by using 300ohm ribbon feeder instead of the 10-way cable. Another constructional method involves the use of individual dipole wires bound together at 2 or 3ft intervals. It takes longer to make up such an antenna from individual wires and the end product is not so neat or clean looking.

If the 3.5MHz band is not required, a useful DX antenna for four bands is possible using just a single mast or support only 25ft (8m) high. A mast of this length can be self-supporting if its base has been well set into the ground. When using such a mast the wires of the inverted-V will actually help to hold it up!

Inverted-V antennas must have a considerable vertically polarised component in their radiation, for their performance when used for inter-UK working on the 28MHz band was much better than that of any horizontal antenna that had been tried at the author's location. It is this vertical radiation which makes inverted-V antennas so useful for long-haul DX work on the 3.5 and 7MHz bands. Unfortunately a 1.8MHz band version is out of the question for most amateurs, for it needs a 90ft (27m) mast or tower!

An ATU *must* be used with this and all types of multiband antennas to ensure the attenuation of harmonics and spurious transmitter outputs. A monoband half-wave dipole can discriminate quite well against its second and other even harmonics, but a multiband inverted-V antenna will radiate them effectively if no ATU is used.

## Half-sized dipoles

There are many occasions when an average amateur discovers that he or she just does not have enough available space to put up a half-wave antenna for one of the LF bands. The author was at one time unfortunate enough to be living in a very restricted urban environment where there was no garden at all. His half-wave wire cut for the 14MHz band ran from the roof ridge of a small terraced cottage to the wall of a house in the next and parallel street behind, and it was suspended about 20ft (6m) above the flat roof of an outbuilding. The 7MHz band was worked by having a quarter-wave top (33ft/10m) with its centre connected to an open-wire feeder. This arrangement worked somehow and contacts were possible, although DX working outside Europe was almost impossible.

The middle section of a dipole does most of the radiating and in fact 71% of the total radiation is from the central half of a dipole's length. Bearing in mind that the author's makeshift dipole's centre 50% was part of the open-wire feeder, it is remarkable that any contacts on 7MHz were possible, for the radiation was being 'cancelled out' by the close proximity in terms of wavelength of the feeder wire pairs.

There are several ways to squeeze a dipole into a garden or other area too short to accommodate a full half-wavelength of wire, but they often involve bending back the dipole ends, dropping its ends, zig-zagging the top or some similar ploy.

By using instead some inductive loading in each dipole leg, it is still possible to achieve resonance and have little actual loss with a dipole only half the normal length for the frequency used. Off-centre loading can be achieved by putting an inductance at any pre-determined position along each of the dipole legs, but there are several considerations to be noted when deciding which is the most useful point for the loading coils.

The greater the distance they are from the antenna centre the more efficient the system will be but, when that distance is increased, larger values of inductance are needed. An increase in inductance results in an increase in ohmic loss, a narrower antenna bandwidth, and a heavy coil which may be difficult to hold up. If a dipole is halved in length (ie the top is only a quarter-wavelength long overall) and a loading inductance is put at the half-way point along each shortened leg, the inductors required must each have a reactance of approximately 950ohms at the operating frequency (Fig 13).

By using this information it is possible to determine the actual inductance needed on each amateur band. Table 3 gives these inductance values, and in addition it gives the frequency tuned by each inductance if it is shunted by a 100pF capacitor. Winding coils to a specific inductance is not an easy task for amateurs for there are so many variables to consider: coil diameter, turns per inch, wire thickness and the coil length-to-diameter ratio. The coils must either be wound on a high-grade insulating material or be stout enough to be self-supporting using thick wire or tubing.

A check of their resonant frequency when in parallel with a close-tolerance 100pF silvered-mica capacitor is not too difficult, however, and they can be trimmed with some accuracy. This can be done by squeezing or opening out the turns of the self-supporting types, or removing or adding turns to those coils which are wound on formers. It is suggested that a dip oscillator coupled to a frequency counter be used when adjusting the coils. This is done without any kind of connection to the antenna wires.

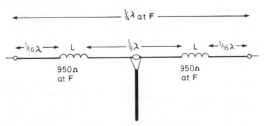

Fig 13. An inductively loaded half-size dipole. The lengths of each wire section may be calculated from the half-wave dimensions given in Table 1. The radiation pattern of a normal full-size dipole will also apply in the case of a shortened version, but its efficiency will be reduced. Coils made from very thick wire or tube, and wound on a high-grade former, will increase the efficiency by reducing ohmic and other losses

An insulator can be inserted half-way along each of the short dipole legs, and the coils (without the 100pF capacitors!) may be soldered into place across them (Fig 14). Those coils wound on formers can be fully weatherproofed by giving them a liberal coating of silicone-rubber sealant. Insulation problems are not great, because the RF voltages will be low where the coils are positioned along the wires at their half-way points.

The finished loaded short dipole may not resonate at exactly the desired frequency at first test, so some adjustments can be made to the lengths of the dipole end sections. A dip oscillator/frequency counter arrangement can be coupled to a single-turn link across the dipole centre (the antenna may be lowered to about 10ft or 3m above ground to do this), or instead a check on the SWR should be made as the transmitted frequency is moved across the band. The lowest SWR normally indicates antenna resonance. An antenna RF noise bridge can also be used to check for dipole resonance when used in conjunction with a receiver.

**Table 3. Loading inductors for half-wave dipoles**

| Band | Inductance | Frequency when tuned with 100pF in parallel |
|---|---|---|
| (MHz) | (μH) | (MHz) |
| 3.6 | 40 | 2.6 |
| 7.0 | 25 | 3.2 |
| 14.0 | 12 | 4.5 |
| 21.0 | 8 | 5.6 |
| 28.0 | 6 | 6.6 |

Even a properly resonated, loaded half-size dipole will not be so effective as a full-sized antenna, but it will still prove to be a very useful radiator and show the same directional characteristics etc. The shortened dipole must not be confused with a trap dipole, a type which is little shorter than a normal half-wave on the lowest band covered. Trapped systems are used to allow multiband operation from one antenna using a single feeder and they are not easy to make and set up correctly.

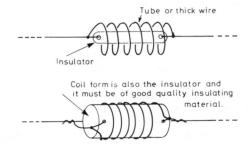

Fig 14. Two ways to insert the loading coils along the dipole wires. Self-supporting inductors can easily be adjusted by squeezing the turns together or apart. The loading coils can be set to the correct inductance before they are used in the antenna, and any final trimming to resonance will only involve the length of the end wires

# Chapter 2

# Centre-fed antennas using tuned feedlines

The half-wave dipoles discussed in Chapter 1, although simple and efficient radiators of RF energy, are very restricting to the amateur who wishes to operate on several of the amateur bands. At best they will be useful on just their fundamental frequency and also on the third harmonic of that frequency, ie a dipole cut for 7MHz will also perform quite well on 21MHz and have an acceptable match to the 75ohm feeder, although on this higher frequency the radiation pattern will change from that of a dipole to that of a 1½-wavelength wire (see Fig 15).

By using a tuned feedline, however, a dipole top of any length can be made to radiate efficiently over a wide range of frequencies. This form of antenna is generally known as a 'doublet' or even as a 'centre-fed Zepp', the latter name really being a contradiction of terms, for a true Zepp is essentially a resonant wire fed at one end.

## Tuned feedlines

Open-wire feedlines have a characteristic impedance which relates to the diameter of the wire used and the spacing between the feed wires. This impedance is important in many applications but is of no consequence when considering centre-fed antennas which use tuned lines. The antennas discussed in this chapter mostly use tuned feedlines and therefore the impedance of such lines can be disregarded.

Tuned feedlines operate on the principle that they are really a part of the antenna and have 'standing waves' along their lengths. Standing waves are a feature of most radiating wires but, if two such wires of equal length are closely spaced (in terms of wavelength) and fed in antiphase, they will radiate just a very small proportion of the RF power applied. A clearer understanding of the action of tuned feeders may be gained from Fig 16.

The instantaneous current along a centre-fed antenna when each leg is ⅝-wavelength long is shown in Fig 16(a). For simplicity the RF generator connects right at the antenna centre. It will be seen that the currents in the half-wave long ends of the antenna are in phase (they seem to run in the same direction) and that the small currents in the two eighth-wave sections near the centre are in opposition

to them. The RF in the smaller centre section will still contribute towards the total radiation from the antenna, but will also have an influence upon its pattern of radiation.

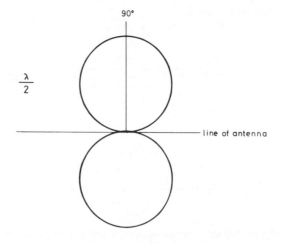

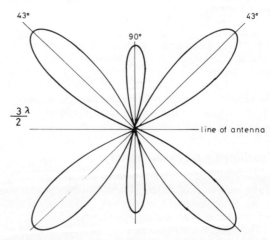

Fig 15. Horizontal polar diagrams for half-wave and 1½-wave horizontal wires

15

This picture changes dramatically when all but the final eighth-wavelengths at the antenna ends is pulled down and runs in parallel towards the RF source (Fig 16(b)). Along the length of the parallel section (really now a feeder) are equal and opposite currents, and therefore this feeder will radiate very little. The total radiation will be from the truncated and tiny top section where the antenna currents are in phase.

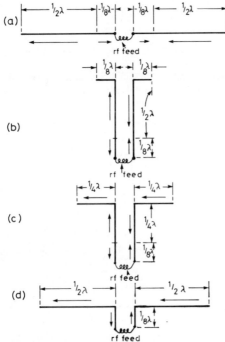

**Fig 16. Four different doublet antenna arrangements using the same total length of wire for the tops and the feedlines. In each example, the antenna currents along the doublet tops are in phase, while the feeder currents along each leg are anti-phase and will cancel**

Fig 16(c) shows an intermediate situation which has a full half-wavelength making up the radiating 'top', and Fig 16(d) illustrates the in-phase currents along a top made from two half-wavelengths. In this diagram the feedline is shown as only one eighth-wavelength long, but it can be of any convenient length without any impairment to the performance of the radiating antenna top.

## Feedlines – physical factors

As mentioned above, the two wires which make up the feeder must be of equal length, and they must in addition be out in the clear, well away from conducting external objects such as pipes and metal guttering.

Their spacing can be remarkably wide before they begin to radiate, and spacings of 6in. (15cm) are in order, even on the 28MHz band. This means that on 1.8MHz the feeder

could have its wires 8ft (2.4m) apart before any significant feeder radiation took place! Standard textbooks on antennas seem to show only pretty feedlines equally spaced along their lengths, but in fact if both the feeder wires are of equal length the spacing may vary. This means that the spreaders which are used to hold the wires apart could, without detriment, be of differing lengths. This may appear heretical, but the fact that all the spacers are equal in length does nothing to enhance the effectiveness of the feeder.

Although the spacing is of little importance, any sharp bends in the run of tuned feeders must be avoided. A length of feeder when turning in towards a shack window must not be bent at an acute angle. Bends of large radius or with angles greater than 90° are the best, and tuned feeders must not run along or be close to walls etc. Untuned or 'flat' feedlines which are correctly matched to their characteristic impedances at either end can be put into plastic or rubber pipes and buried with impunity, but tuned lines must be out in the open.

The use of transmission-line (quarter-wave transformers) is described in Chapter 6.

## Feeder losses

The losses in tuned lines include any loss by radiation (which is normally very small), and any ohmic loss in the conductor wires. Assuming that feedlines are fabricated from 18 to 14SWG copper wires, the ohmic losses are insignificant and those that exist are mostly related to the dielectric properties of the material used for the spacers.

At UHF or VHF spacers are seldom used, the wires being held taut and parallel with just air as the dielectric. Open-wire matched lines with spacers arranged at 18in. (45cm) intervals will show an attenuation of 0.03dB per 100ft (30.5m) at 3.5MHz, rising to 0.25dB at 144MHz. This means that a 3dB power loss (half power) will only occur when the feedline is 10,000ft (3km) long at a frequency of 3.5MHz! Tuned lines will exhibit higher losses.

For convenience, 300ohm ribbon feeder may be used as tuned line, especially the Swedish slotted variety. The finest available ribbon feeder will have a greater loss than open-wire line, but this will still be relatively insignificant. The older type of flat 300ohm ribbon feeder, when used as a matched line, has an attenuation of 0.18dB per 100ft at 3.5MHz, rising to 1.55dB at 144MHz. However, this older ribbon detunes badly in wet weather. The slotted variety has a better performance and its 'semi-air' spacing and water-shedding characteristics make it ideal. Its use also avoids the tedium of making up a long run of open-wire feeder.

## The velocity factor of tuned lines

The inherent self-inductance and capacitance distributed along the length of a feeder gives rise to its 'velocity factor'. The dielectric material between the wires also plays a large part in this factor. The effect is to slow down slightly the wave on the line, and the electrical length of a feedline is therefore somewhat less than the calculated resonant length.

The velocity factor is only of significance in certain antenna types, for example those requiring quarter-wave 'stubs'. Such stubs are needed for the four-element collinear antenna illustrated in Fig 21 and described later. Open-wire lines have a velocity factor of about 0.975 which means that a quarter-wavelength stub at 7MHz will be 10in. (25cm) shorter than a basic electrical quarter-wavelength.

## Reactance

Tuned feeders can exhibit reactance at their fed point, and this reactance may be either inductive or capacitive depending upon the frequency, the feeder length and the length of the antenna top. With any given antenna which uses tuned feeders, the reactance will be different on each frequency band, and it can be that on one or more bands there will only be resistive impedance which is much easier to cope with when using an antenna tuning unit (ATU).

The use of an ATU, also known as an 'ASTU' (antenna system tuning unit) or 'AMU' (antenna matching unit), is essential when using an antenna with a tuned feedline. In most cases the ATU will be able to 'tune out' the reactance present, but unfortunately there is no ATU design which will cope with an infinite range of impedance or reactance, so in practice certain combinations of antenna and feeder lengths must be avoided. An antenna which appears almost impossible to match on just one amateur band may have this corrected by the addition or subtraction of feet, or tens of feet, of feedline.

Louis Varney, G5RV, has suggested that electrical lengths which are odd multiples of a eighth-wavelength (this includes one leg of the top and the feeder length in the case of the doublet and similar antennas) should be avoided as these are the most likely to give severe reactance problems. Table 4 lists the different lengths to avoid for the amateur bands and will be helpful when designing a centre-fed antenna for multiband use.

## The basic doublet antenna

This antenna type is perhaps the most useful simple multiband antenna for amateur use as it can be located away from the house and/or TV receivers and other devices which may be prone to EMC problems. It does not need any special earthing or counterpoise arrangements either, but it is an antenna to be avoided by the operator who wishes to connect an antenna directly to the transceiver.

An ATU is essential when using a doublet. Some

authors have suggested that a 4:1 step-down balun can be connected at the bottom end of tuned feeders to allow a direct connection to the low-impedance input socket of a transceiver. This advice is most unwise, for the reactance present on open-wire feeders will almost certainly give rise to the overheating and eventual destruction of the balun on one or more bands. A balun must only be used when a *correctly matched* feedline of a nominal characteristic impedance is used, such as in the feeder (300ohms) from a folded dipole, and *never* with a tuned feeder which has standing waves.

**Table 4. Lengths to avoid when designing multiband doublets with tuned feeders**

| Band (MHz) | Lengths (half the top plus feeder length) | | | | | |
|---|---|---|---|---|---|---|
| 1.8/1.9 | 185'<br>56.38m | 307'<br>93.7m | 430'<br>131m | | | |
| 3.6 | 96'<br>29.26m | 160'<br>48.76m | 224'<br>68.27m | | | |
| 7 | 49.5'<br>15m | 82.5'<br>25.14m | 115.5'<br>35.2m | 148.5'<br>45.26m | | |
| 10.1 | 34.5'<br>10.5m | 57.5'<br>17.52m | 80.5'<br>24.53m | 103.5'<br>31.54m | | |
| 14.15 | 24.75'<br>7.54m | 41.25'<br>12.57m | 57.75'<br>17.6m | 79.25'<br>24.15m | 90.75'<br>27.66m | 107.25'<br>32.68 m |
| 18.1 | 19.5'<br>5.94m | 32.5'<br>9.9m | 45.5'<br>13.86m | 58.5'<br>17.83m | 71.5'<br>21.79m | 84.5'<br>25.75m |
|  | 97.5'<br>29.71m | 110.5'<br>33.68m | | | | |
| 21.2 | 16.25'<br>4.95m | 27'<br>8.22m | 38'<br>11.58m | 48.75'<br>14.85m | 59.5'<br>18.13m | 70.5'<br>21.48m |
|  | 81'<br>24.68m | 92'<br>28.04m | 103'<br>31.39m | 114'<br>34.74m | | |
| 24.94 | 14'<br>4.26m | 23.25'<br>7.11m | 32.75'<br>9.98m | 42'<br>12.8m | 51.25'<br>15.64m | 60.75'<br>18.51m |
|  | 70'<br>21.33m | 79.25'<br>24.17m | 88.75'<br>27.05m | 98'<br>29.87m | | |
| 29 | 12'<br>3.65m | 20'<br>6.09m | 28'<br>8.53m | 36'<br>10.97m | 44'<br>13.41m | 52'<br>15.84m |
|  | 60'<br>18.28m | 68'<br>20.72m | 76'<br>23.16m | 84'<br>25.6m | 92'<br>28.04m | 100'<br>30.48m |

The arrangement shown in Fig 17 is very simple, the only critical factor being the avoidance of certain combinations of feeder length/top leg length as shown in Table 4. The antenna is essentially a balanced system and each half of the top plus each wire in the feedline must be equal in length. Referring to Fig 17, AB = CD and BE = CF. The antenna top is not cut to resonate at any particular frequency (unlike the half-wave dipole), and any length may be chosen to suit an individual location.

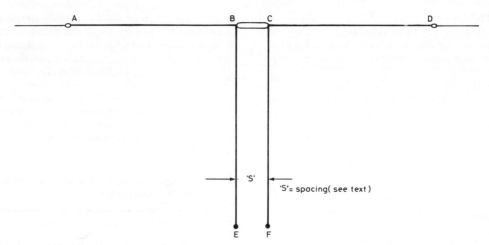

Fig 17. A basic doublet antenna, showing its inherent balance. The two legs of the top AB and CD are equal in length, as are the feeder wires BE and CF

A minimum top length of a quarter-wavelength (an eighth-wavelength for each leg) at the lowest contemplated operating frequency is suggested. Even when the top of the doublet antenna is a quarter-wavelength long, the antenna will still be an effective radiator. The 'missing' quarter-wavelength is taken up by the upper part of the feeder, so the antenna current in this section will cancel and not be radiated. This reduces the overall efficiency of the antenna but it will still tune up easily.

A similar shortened antenna was used by the author 40 years ago when located in a cramped and gardenless urban location. The maximum antenna top length there was restricted to 30ft (9m) but, when using a transmitter power output of 75W on 7MHz, consistently good reports were received from all Europe and even the USA. This diminutive antenna was strung out between a couple of chimney stacks and was only about 25ft (8m) from the ground.

## Radiation patterns

Where a total top length of 100ft (30.5m) is possible, a doublet antenna will work well on all the bands 3.5MHz to 28MHz and, if the feeders are strapped at their lower end and tuned against ground, it will also be effective on 1.8MHz. The radiation pattern in the horizontal plane will be similar to that of a half-wave dipole when it is used on 3.5MHz; as two half-waves in phase on 7MHz; as a 1½-wavelength antenna on 14MHz with a six-lobed pattern of radiation; and more and more like a long wire on the higher-frequency bands.

The 14MHz pattern can be seen in Fig 18 where it has been superimposed upon a crude great circle map centred on Britain. It shows that a good world coverage is possible by arranging an approximate east/west run for the antenna top. If the wire direction ran from northwest to southeast the major lobes would then include Australia, South Africa and Central America.

Of course, the expected radiation pattern will not be obtained if the antenna top is bent. Also, to achieve the low-angle vertical radiation needed for DX working, the antenna must be at least one half-wavelength above ground at the operating frequency.

On 14MHz this means a height of at least 30ft (9m). On 3.5MHz an antenna height of 60ft (18m) still only represents a quarter-wavelength, and most of the radiation will therefore be at high angles. This circumstance is, however, quite suitable for UK and European contacts, and signal strengths will normally be high.

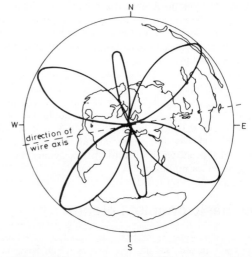

Fig 18. The horizontal radiation pattern of a 1½-wavelength doublet antenna which is at least a half-wavelength above the ground. The crude great circle map demonstrates how wide coverage can be achieved with such a simple antenna

Table 4 shows that when using doublet legs of 50ft (15.2m) together with 54ft (16.4m) of feeder there ought to be little difficulty with reactance on most amateur bands. There are of course many other combinations of top length and feeder length which can be chosen, either the feeder or the top being adjusted in length to suit individual locations.

## Two half-waves in phase

If the top of a doublet is made from a pair of half-wavelength wires the antenna then becomes a simple collinear with some gain over a dipole. The radiation pattern will be similar to that of a dipole but it will more nearly resemble a narrowed '8' and have a reduced radiation towards the ends of the wires. Its theoretical gain over a half-wave dipole will be 1.9dB. To achieve a greater gain when using two half-waves in phase, the spacing between the adjacent ends of the two half-wavelength wires must be about 0.45 wavelength. This spacing is easy to arrange when using separately fed, individual half-wave dipoles in phase, and then the gain over a single-dipole element becomes 3.3dB. To do this involves a pair of feedlines equal in length and a considerable space (at least 100ft or 30m on 14MHz). Fortunately there is a very effective and simpler substitute which has almost the same antenna gain; this antenna is known as the 'extended double Zepp'.

## The extended double Zepp

By the simple expedient of lengthening both the doublet elements of an antenna consisting of two half-waves in phase, the effective spacing between the inner ends of the half-wave sections becomes greater. In this way, with an antenna top made from two 0.64 wavelengths, the effective spacing between the two half-waves becomes 0.28 wavelength.

This spacing gives a gain of about 3dB over a single half-wave dipole antenna. A gain of 3dB is equivalent to a doubling of the transmitted power and it is a very worthwhile feature of this antenna. Fig 19 shows the extended double Zepp arrangement, and it will be noticed that the antenna currents in the added sections B and C, although in phase with each other, are opposed to the currents in the two longer half-wave sections of the top.

The power radiated by the centre sections contributes

towards four weak lobes, each of which lies approximately 35° from the line of the wire top (see Fig 20). The dimensions for extended double Zepp antennas are given in Table 5.

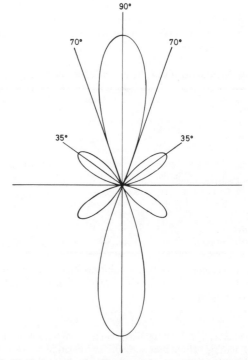

Fig 20. Horizontal polar diagram of the extended double Zepp

An added bonus to be gained with this type of antenna is that it will also radiate well on other bands, but not of course as a two-element collinear. The 14MHz version will have a total top length of about 83ft (25.29m) and will behave as a 'long dipole' on 7MHz. Even on 3.5MHz, where each leg is about 25ft (8m) short of a quarter-wavelength, the antenna will still radiate effectively.

### Reflectors for the extended double Zepp

If a pair of parasitic wire reflector elements is located

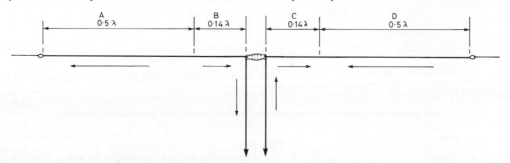

Fig 19. The design criteria for extended double Zepp antennas

behind the half-wave sections of the extended double Zepp antenna, the radiation will then be mostly in one direction and the antenna gain will increase by almost another 3dB. This will give a total theoretical gain over a half-wave dipole of about 6dB (four times effective power). Such an arrangement requires two extra antenna supports and takes up more space but it would be a very useful antenna for consistent and regular DX working towards a particular country or station.

**Table 5. Dimensions of the extended double Zepp**

| Frequency (MHz) | 0.5 wave | 0.14 wave | Leg length (A+B) |
|---|---|---|---|
| 7 | 66' 9" 20.34m | 18' 6" 5.63m | 85' 32" 26m |
| 14.2 | 32' 8" 9.98m | 9' 2.74m | 41' 8" 12.64m |
| 21.2 | 22' 6.7m | 6' 1.82m | 28' 8.53m |
| 28.5 | 16' 5" 5m | 4' 6" 1.37m | 21' 6.4m |

The spacing between the collinear elements and the reflectors should be about 0.15 wavelength, which is 9ft 9in. (2.97m) on 14MHz, 6ft 6in. (1.98m) on 21MHz and 4ft 10in. (1.46m) on 28MHz. The reflectors must be 35ft 2in. (10.71m), 23ft 6in. (7.16m) and 17ft 6in. (5.33m) long respectively for the three wavebands. A nylon or similar cord running behind the centre 0.28 wavelength section of the double Zepp can be used to join and hold up the reflectors at their inner ends.

## Four collinear elements

The antenna illustrated in Fig 21 is a particularly useful one for an operator who needs a bidirectional beam (fixed) which has a gain of about 4.3dB over a dipole. It is designed for the 28MHz band and its total length of u. der 70ft (21m) is not excessive for average garden plots. It only has to be 20ft (6m) from the ground to perform well on its design frequency but would, like most antennas, be even more effective at greater heights, where the screening effects of buildings and trees etc are reduced. The use of two quarter-wave stubs adds to the total wire length and this makes each leg about 50ft (15.2m) long. On 3.5MHz the antenna is a 'short dipole', with its maximum current at the 'dipole centre' some 16ft (5m) down the feedline. Being a low dipole on this band, it is only really useful as a high-angle radiator for semi-local working, say, between 100 and 600 miles range.

When used on its design frequency the currents in the half-wave elements are kept in phase by using quarter-wave shorted stubs. These are each 8ft (2.43m) long and their length was calculated bearing in mind the velocity factor of the wire pairs which make up each stub. If 300ohm ribbon is used for the stubs, another velocity factor of either 0.82 for the older flat ribbon or 0.87 for the slotted Bofa variety must be kept in mind. In the latter case the stubs must each be 7ft (2.13m) long.

To prevent undue waving about in the slightest breeze, it is recommended that small weights made from pebbles should be taped at the bottom of each stub. If you live in the Shetlands it might be as well to tie a thin cord to each stub and take it down to a suitable anchor point on the ground below! The RF currents in the stubs cancel and therefore they do not radiate.

## The G5RV antenna

Modern all-solidstate amateur-band transceivers, using output stages that are easily damaged when operated with

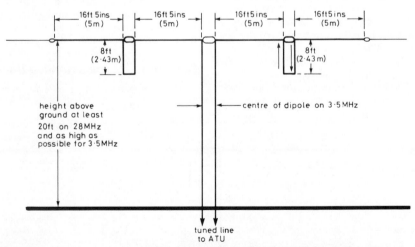

Fig 21. The measurements for a four-element collinear beam antenna designed for the 28MHz band which will also be a useful high-angle radiator on 3.5MHz

high SWR on the feed cable to the antenna or with special ALC circuitry to run down power in such situations, have induced many amateurs, especially newcomers to the hobby, to look for antenna systems which can connect directly to their equipment. Simple half-wave dipoles, trapped dipoles, verticals, and multi-element beams often have coaxial feeders with an impedance of 50ohms which can of course be connected directly to most transceivers without any ATU. There is no way, however, that a centre-fed multiband doublet of the type previously discussed can be used without an ATU, for the tuned feeders will present a wide range of impedances and reactances on the different amateur bands.

Louis Varney, G5RV, designed his now-famous 'G5RV' antenna as long ago as 1946, and many of today's 'black box' owners have opted to use his antenna design with its multiband option and coaxial (or twin-lead) cable feeder. Unfortunately the results obtained with this antenna have often not been up to expectations on more than two bands. Louis Varney updated his antenna in the July 1984 issue of *Radio Communication* (RSGB), and also importantly stressed the need for an ATU between the coaxial feedline and the transceiver.

Contrary to the design logic of most multiband antennas, the G5RV was not designed to be a half-wave dipole on its lowest operating frequency (usually 3.5MHz) but instead a 1¹/₂-wave centre-fed doublet on 14.15MHz. The total length of the G5RV antenna top is 102ft (31.27m) made from two equal 51ft (15.54m) lengths (see Fig 22).

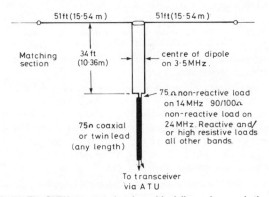

**Fig 22. The G5RV antenna, showing critical dimensions and other details**

Instead of simply bringing down a tuned line or ribbon feeder, G5RV arranged that there should be a 34ft (10.36m) matching section of open-wire feeder, which had connected to its lower end a length of either 75ohm impedance twin lead or 80ohm coaxial cable (see Fig 22). If 300ohm ribbon is used for the matching section, the old type must be cut to a length of 28ft (8.5m) and the slotted variety to a length of 30ft 7in. (9.3m), which takes into account the different velocity factors of the two ribbons. Unfortunately the match to the coaxial or twin feeder at its junction with

the lower end of the matching section is only good on 14MHz and 24MHz. If 50ohm coaxial is used the VSWR on these bands will rise to 1.8:1.

G5RV describes in some detail the differing loads presented to a 75ohm feeder (flat twin or coaxial) at the point where it joins with the matching section. A brief outline of his findings is as follows:

| | |
|---|---|
| 3.5MHz | reactive load |
| 7MHz | reactive load |
| 10MHz | reactive load |
| 14MHz | resistive load, approx 90ohms |
| 18MHz | high-impedance load, slightly reactive |
| 21MHz | high-impedance load (resistive) |
| 24MHz | resistive load approx 90/100ohms |
| 29MHz | high-impedance load, slightly reactive |

From the above it becomes obvious that it is *essential* to use an ATU between the low-impedance feeder and the transceiver. Although the equipment will then 'see' a correct matching impedance of about 50ohms there will be a considerable mismatch in the system and standing waves along the feeder. Towards the end of the description of his antenna, G5RV mentions that the most efficient feeder to use is the open-wire variety, all the way down from the centre of the antenna to the equipment, in conjunction with a suitable ATU for matching. He added that by using 84ft (25.6m) of open-wire feeder the system will permit parallel tuning of the ATU on all bands.

## Practical considerations

Many amateurs use the G5RV antenna with considerable success, but the author prefers the use of either open-wire or 300ohm ribbon to feed the horizontal top. With an ATU such a feed will result in high-performance, all-band working. The G5RV restricts efficient working to a couple of bands and, even on those bands where the match to the open-wire line section is reasonable, an ATU is still needed at the transceiver end of the low-impedance feedline. The basic doublet has no critical dimensions; its only strict criterion is balance, with both top sections being equal. A G5RV must have a top length of 102ft overall, whereas the doublet can be put up in restricted locations where perhaps it is only possible to run out 90ft of wire or less.

If coaxial feeder is used, its weight will pull its centre down considerably, and it may be necessary to use a centre support and slope the antenna down on each side, so making an inverted-V. Doing this will alter the expected radiation patterns and it will also detune the antenna. ZS6BKW found that one effect of converting his antenna (see below) into an inverted-V was to change its resonant frequency. When being used on 14MHz, its frequency was lowered by 50kHz (which is a small change in frequency, only 0.3%) when the angle between the wires was decreased to 120°. This came down another 50kHz when the angle was sharpened to 85° (much too acute an angle for an inverted-V anyway).

Any basic doublet antenna can be set up in the inverted-V configuration, and any detuning effects of the angle between the wires will be taken up by the ATU. Top resonance is not a feature required for the proper working of an elementary doublet antenna, so dropping its ends does not cause problems. If the current maximum is near or at the antenna centre, this antenna will be very effective. Where space is very limited, the ends of the top wires may be dropped down vertically, but each half of the antenna must remain a 'mirror image' of the other, again to ensure balance. In really desperate situations the two top sections can be bent in the horizontal plane, but this can unbalance the antenna and the feed wires will radiate. Such an imbalance is also more likely to induce or worsen EMC problems, but there are some very restricted locations where this risk must be taken.

If a coaxial feeder is being used correctly as a 'flat' untuned line its characteristic impedance will be matched at each of its ends. If this is the case then the coaxial cable may be safely buried along its run with no detriment to its operation. This can be useful when an antenna is located at a considerable distance from an operating position, but unfortunately the coaxial cable feed of a G5RV antenna must never be buried. The feeder is not correctly terminated and operates with standing waves along its length. It must thus be kept well away from metal and large objects, and most certainly not buried!

An examination of the G5RV antenna reveals that there is little advantage in using this design, and the author would always instead advise the use of tuned feeders all the way from the antenna centre down to an ATU. Doing this will ensure that most of the RF power being put into the feedline will be radiated on all the bands used.

## The ZS6BKW computed design

ZS6BKW developed a computer program to determine the most advantageous length and impedance of the matching section and the top length of a G5RV-type antenna. He arranged that his antenna should match as closely as possible into standard 50ohm coaxial cable and so be more useful to the user of modern equipment. His findings show that a good match with low SWR readings can be obtained on five of the amateur bands.

The G5RV antenna total top length of 31.1m was reduced to 27.9m, and the matching section was increased from 10.37 multiplied by the velocity factor to 13.6 multiplied by the latter. This matching section *must* have a characteristic impedance of 400ohms which can be made up from a specially made length of open-wire line. A pair of 18SWG wires spaced at 1in. (2.5cm) apart will be suitable. ZS6BKW's antenna gave the following SWR figures:

| | | | |
|---|---|---|---|
| 3.65MHz | – | 11.8:1 | poor |
| 7MHz | – | 1.8:1 | good |
| 10MHz | – | 88:1 | very poor |
| 14MHz | – | 1.3:1 | good |
| 18MHz | – | 1.6:1 | good |
| 21.2MHz | – | 67:1 | very poor |
| 24MHz | – | 1.9:1 | fairly good |
| 29MHz | – | 1.8:1 | good |

The ZS6BKW antenna has reasonable SWR figures on five bands, which is an improvement on the original G5RV, but they are still not good enough to use the antenna without an ATU for correct matching.

## Some practical considerations

It may be an illusion but it seems that the British Isles are becoming more windy each year. This may be due to the long-term but gradual changes in the atmospheric pressure which have been noted by weather pundits, but personal experience has convinced the author that there are now fewer really calm, windless days in each year. This means that any outdoor wire-antenna system will swing about for much of the time and, unless sensible precautions are taken when constructing it, metal fatigue will take its inevitable toll.

Multi-strand, plastic-covered wire is cheap and easily found on the 'surplus' market, and there is a temptation to use this kind of wire for most antenna work. These wires when used will have a useful life not normally longer than a couple of years. The constant movement of the antenna results in the breaking of the conductors, leaving the plastic covering undamaged. When this happens (it has to the author and several of his 'locals') the actual break is difficult to locate and repair.

Single-strand, hard-drawn copper wire of 18 or 16SWG is much better for all antenna work, including particularly the fabrication of open-wire lines. It may seem to be more expensive as an initial outlay, but these wires will last for many years. There are some lengths of copper wire 'resting' in the author's wire box which have been part of a variety of antennas during the past 30 years.

It must be mentioned that 75ohm twin feed is particularly prone to internal breaks if it is allowed to swing freely.

### The centre connection

The usual breaking point along amateur antennas is where the feeder connects to the radiating section. A feeder naturally hangs down and swings in the slightest breeze, inducing stress at the actual connection point.

The most common arrangement used by amateurs is shown in Fig 23(a). Here the centre insulator is a single Pyrex glass or glazed ceramic unit to which the antenna wires are joined. The wire ends are usually twisted and

(a)

(b)

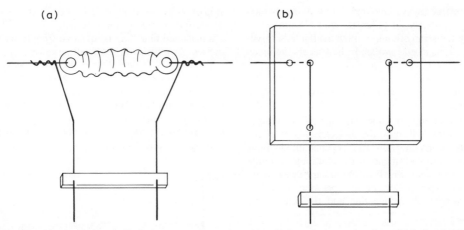

Fig 23. (a) Doublet-to-feeder connections with a twisted and usually soldered join at the centre insulator. (b) An insulated centre block which allows the use of unbroken wire lengths

soldered, then the two feeder wires are twisted on top and they too are soldered.

Solder soon weathers and copper wire rapidly acquires an oxide patina. When there are dissimilar metals (lead/tin and copper), a metallic oxide and current flow, problems will arise! The resistance of the connection can increase and reduce the power in the top wires when there is a large RF current at that point.

Even worse is the so-called 'rusty bolt effect'. The corroded joint behaves like a semiconductor diode which, when energised by the transmitted power, can generate harmonics across an enormous range of frequencies right up to UHF. This can give rise to EMC problems and is difficult to recognise and put right.

A centre block, made from a piece of ¹/₄in. or ¹/₂in. (6 or 12mm) thick insulating material which can be easily drilled (see Fig 23(b)), will allow the use of just two single lengths of wire to make up the top sections and the tuned feeder, without any soldering or twisted joins. If the block is made from Perspex or a similar material which has a polished surface it will be ideal, but it must be mentioned that ¹/₄in. Perspex may crack if subjected to the strain of a long antenna.

300ohm ribbon feeders present a greater problem because their multi-strand conductors are not robust and will soon break under strain or repetitive bending. Fig 24 shows how to make up a centre block which can alleviate such difficulties. Soldered connections have to be made to the inner ends of the top wires but there is no physical strain on these. The weight of the feeder is taken away from the connection by the way it is threaded through slots in the insulating block. A tightly knotted nylon cord (heavy mono-filament fishing line) near the base of the block gives an additional restriction to movement of the ribbon.

## Open-wire feeder construction

There was once a time when it was possible to buy ready-made open-wire feedline. Today, such feeder would be prohibitively expensive, largely because of the labour involved in its construction, and it falls upon the amateur to 'roll his own'. Most articles dealing with antenna systems using open-wire feeder carry advice on the making of such lines.

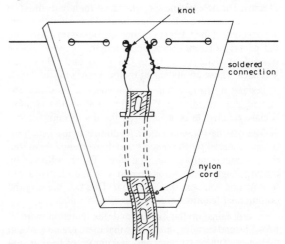

Fig 24. A centre block suitable for the connection of a 300ohm impedance ribbon to an antenna centre

The spacing spreaders always seem to present the greatest challenge to the ingenuity of the authors, and these may range from plastic hair curlers to sections of plastic milk bottles! In the 'pre-plastic' age, before the second world war, spreaders were often made from wooden dowel which had to be boiled in paraffin wax. There were also glazed ceramic spreaders available but these items were heavy and expensive. It is still possible to buy ready-made spacers and there are at least two advertisers of these. For a small expenditure and not too much effort feeder spacers can easily be home-made.

The author has always favoured Perspex strips which measure approximately 5in. (13cm) by ¾in. (18mm) wide, and cut from the scrap offcuts so often available at 'give-away' prices at local glassworks. If the spacers are made short, say 1 to 1½in. (25 to 30mm) long, the length of the leakage path between the wires is greatly reduced, and any accumulation of dust, soot and grime will give problems in damp weather.

Using very thin wire will mean the use of many more spreaders, and this will raise the dielectric losses of the line. A sturdy 18 or, better still, 16SWG enamelled wire is ideal although, as already mentioned, a continuation of the actual antenna top wires which may be of hard-drawn, bare copper is very satisfactory. Fewer spreaders will also reduce the weight of the feedline and its pull on the antenna.

## Making the feeder

Two equal lengths of wire can be tied to an outside feature such as a fence post or railing, and then pulled out tightly towards another tie point. The spacers can then all be threaded on to the wires, after which the wire ends are secured to the second tie point. The spacers must then be equally spaced along the feeder at intervals of about 18in. (45cm). The holes in the spreaders near their ends must be just large enough to allow the wires to be pushed through. To ensure that the feeders will not slip out of position with the passage of time, some short lengths of wire can be twisted on to the feeder wires just above and below each spreader. These are indicated by the arrows in Fig 25(a).

Instead of drilling holes in the spacers, they may be slotted as shown in Fig 25(b). This method allows additional spreaders to be added after the line is completed.

The very best way to make a feedline is to use a thermo-plastic material for the spacers and then arrange them into their correct positions along the wires. This is followed by putting a high current through each wire in turn, in order to heat up and bond the spreaders to it. This was the way the commercial feedlines were made 30 years ago.

Avoid joins in the feeder and the antenna wire if possible, and certainly make sure that there are no kinks in the wires. The preparation and making up of feeders and doublet wires cannot be done satisfactorily indoors and a fine day is an important factor!

It is often sensible to anchor an open-wire feedline to stop excessive swaying, and this can be done with the help of the nylon fishing line referred to earlier. This line is often stained blue, and is invisible from more than a few metres. Any bends in the feedline must be of as large a radius as possible, and there must be no sudden and sharp bends. The line must be kept as far as possible from walls, down-pipe and gutterings etc, and it must also drop down from the antenna centre at a right angle for at least a quarter-wavelength at the lowest frequency to be used. If the feeder runs under one leg of the antenna top the system will become unbalanced.

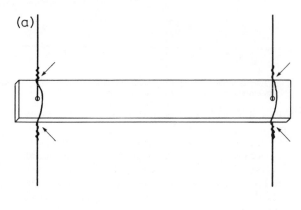

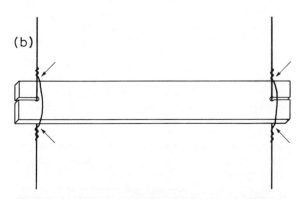

**Fig 25. (a) An open-wire feeder line spacer with drilled holes for the wires, and wire 'retainers' to prevent slip. Spacers of this type have to be threaded onto the feeder wires before they are set into their final positions. (b) A spacer with slotted ends which can be easily fixed at any point along a completed length of feeder**

Chapter 3

# End-fed long wires

A single length of wire fed from one end is perhaps the simplest antenna type available to the radio amateur but, like most things in this world, all is not quite so simple as it first seems! Unterminated end-fed wires which are longer than an electrical half-wavelength will operate as Hertzian antennas, and not as Marconi systems which must be tuned against ground.

This Hertzian concept seems to befog the minds of some end-fed wire users for, when thinking of 'antenna resonance', these folk carefully calculate and then cut their wires to exact resonant lengths. This practice is quite in order if the outdated Zepp feed is to be arranged at one end of the wire but, when using just a single wire which runs right to the ATU without any kind of feedline, the actual wire length is almost immaterial.

The Zepp feed method cannot be recommended because it is an inherently unbalanced system. Equal currents in its open-wire feeder can only be obtained when the flat-top radiating antenna section is an exact and resonant half-wavelength long. On all other frequencies the currents in the feeder will not be equal.

A professional antenna designer (the well-known amateur 'Dud' Charman, G6CJ) discovered many years ago that some of his antenna designs were failures because he had been using the standard Zepp feed system. He suggested a way to overcome the method's shortcomings in an article published in the *RSGB Bulletin* of December 1955. His modified Zepp feed is, however, not suitable for multiband operation and is also rather complex. It is more useful to the commercial user for fixed-frequency work.

## Wire length and impedance

It has often been said that "Nature abhors a vacuum", and it can equally be stated that "The far end of a single-wire antenna abhors a low impedance"! All wires will display, at any frequency, a high impedance at their far ends, and this fact makes it a simple matter to work forwards along the wires in half-wavelengths or quarter-wavelengths to determine the approximate impedance at their feedpoints.

Some impedances are difficult to match with an ATU, particularly those which are very high or low, so it is prudent to arrange that such conditions are not present at the ATU end of an end-fed wire. Lengths of wire which are close to odd multiples of a quarter-wavelength (or just less than this) are particularly bad, and the latter lengths will display a capacitive reactance which will need to be tuned out. When a long end-fed wire is used as a multiband antenna it is almost impossible to determine a length which will avoid some reactance on one or more bands, but this reactance can be quite easily tuned out by using either an inductor (for capacitive reactances) or a capacitor (for inductive reactances) between the wire end and the ATU (see Fig 28).

## Some advantages of long-wire antennas

Unlike the collinear antennas (Chapter 2), which have their half-wavelengths fed in such a way as to ensure that the antenna currents are all in phase, a single-wire antenna which is two half-wavelengths or more long will have alternate half-wavelengths out of phase. When wires are very long in terms of wavelength, the half-wavelengths along their length will also have different amplitudes of current, these currents diminishing along the wire.

Long wires produce quite complex multi-lobe radiation patterns, unlike the simple radiation lobes produced by collinear antenna systems. The longer the wire is made in terms of wavelength the more likely is the radiation to be from off its ends. There will also be an increasing number of minor lobes in other directions as the wire length becomes greater. Fig 26(a) shows the horizontal radiation pattern for a horizontal wire, two wavelengths long. The four main lobes are marked A, B, C and D, and their effective power is about 5dB greater than the radiation from the four minor lobes.

When the wire length is increased to five wavelengths (Fig 26(b)) there are still four main lobes but these are aligned closer to the line of the antenna wire (each being 22° from this line). There are additionally 16 minor lobes, the weaker of which are almost 10dB down from the main lobes. In both these examples, and with all long-wire antennas, there are deep nulls at right angles to the wire.

If a long-wire antenna is at least half a wavelength above the ground at its lowest DX frequency (this is usually 14MHz), the angles of radiation in the vertical plane will be low, and usually between 10 and 15° above the horizon. A single wire which is five wavelengths long will have a power gain over a dipole (22° from the line of the long wire) of more than two times (4dB). This wire, when used on the 28MHz band, will become 10 wavelengths long and will show a quite considerable gain of 7.4dB. The maximum gain will then be in directions closer to the run of the wire.

Wires which are 10 or more wavelengths long will radiate mostly from their ends and they can be aligned to 'point' towards the preferred direction. In terms of decibel gain for money spent, a long single-wire antenna has an advantage over a multi-element Yagi beam, but of course it cannot be rotated! To achieve world coverage, the long-wire enthusiast will need several wires running in different directions and a very large area of ground.

## Practical end-fed wires

Two physical arrangements for end-fed wire antennas are shown in Fig 27. The wire in Fig 27(a) is virtually horizontal along its length, and will show little if any radiation which is vertically polarised. The author used a similar antenna for a number of years when his shack was up near the top of a tall Victorian villa which had a garden sloping away steeply at the rear. The wire ran out horizontally from the top of the window to almost the topmost boughs of a mighty and ancient oak tree in a neighbour's garden about 200ft (61m) away. This wire gave the ex-

pected pronounced nulls at right angles to the wire, and DX contacts in those two directions were almost impossible.

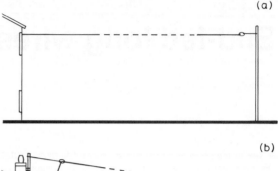

(a)

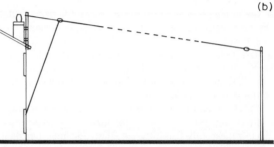

(b)

Fig 27. (a) A horizontal end-fed wire which will have little radiation in the vertical plane. On its harmonic frequencies there will be pronounced nulls at right angles to the wire. (b) A similar wire but the sloping section will allow some vertically polarised radiation and tend to spoil the expected radiation pattern of a long wire operating on its harmonic frequencies. The nulls at right angles to the wire will not be so pronounced

In Fig 27(b) the wire slopes up from a downstairs window to a high point and then runs down at a small angle

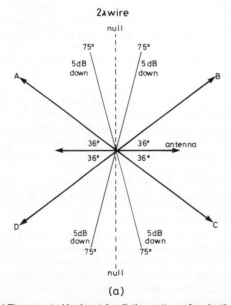

(a)

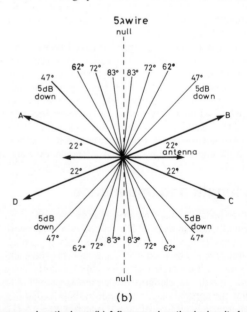

(b)

Fig 26. (a) The expected horizontal radiation pattern of a wire that is two wavelengths long. (b) A five-wavelength wire has its four main lobes closer to the line of the wire and has more minor radiation lobes to the end support. If the near-vertical part of such an antenna is just a small proportion (say one-tenth) of the main top length it will have little effect upon performance, although there must be some radiation (vertically polarised) from the descending 'feed' wire. This vertically polarised section will fill in the nulls in the radiation pattern but additionally may induce some TVI or other EMC problems. The gradually sloping top will lower the angle of radiation towards the far end of the antenna and most of the radiation will also be in that direction.

Long wires which are 30 to 40ft (9 to 12m) above ground will be very effective on 14MHz and the other bands higher than this frequency. Their DX performance on the higher bands will not, however, be matched on lower bands such as 7 and 3.5MHz. Here they will have much higher angles of radiation and will therefore be more suited for medium-distance and short-haul communication.

## Long wire lengths

The individual wavelengths away from the ends of a long wire do not join on to insulators, so their calculated lengths do not have to take into consideration the 'end effect'. They will therefore be closer to the theoretical free-space lengths. The lengths in feet and metres of wires up to 10 wavelengths long on the 14, 21 and 28MHz bands are given in Table 6. On other frequencies the lengths can be calculated by using the expression:

$$\text{Length (feet)} = \frac{984 (N - 0.025)}{\text{Freq (MHz)}}$$

where $N$ is the number of full wavelengths in the antenna. To calculate the lengths in metres use:

$$\text{Length (metres)} = \frac{300 (N - 0.025)}{\text{Freq (MHz)}}$$

A long wire must be at least two wavelengths long to show noticeable gain (about 1.3dB) over a half-wave dipole, and until it is four wavelengths long its gain remains small. It

### Table 6. Resonant lengths of long wires

| Wave-lengths | 14.15MHz | 21.2MHz | 28.5MHz |
|---|---|---|---|
| 1 | 67' 9" (20.66m) | 45' 3" (13.79m) | 33' 8" (10.25m) |
| 2 | 137' 4" (41.86m) | 91' 8" (27.9m) | 68' 2" (20.78m) |
| 3 | 206' 10" (63.0m) | 138' 0" (42m) | 102' 8" (31.3m) |
| 4 | 276' 5" (84.25m) | 184' 6" (56.2m) | 137' 3" (41.83m) |
| 5 | 345' 11" (105.4m) | 230' 11" (70.37m) | 171' 9" (52.35m) |
| 6 | 415' 6" (126.6m) | 277' 4" (84.5m) | 206' 4" (62.87m) |
| 7 | 485' 0" (147.8m) | 323' 9" (98.67m) | 240' 9" (73.4m) |
| 8 | 554' 7" (169m) | 370' 2" (112.8m) | 275' 4" (83.9m) |
| 10 | 693' 8" (211.4m) | 462' 6" (140.9m) | 344' 5" (104.9m) |

is only when eight or more wavelengths are used that the power gain becomes really significant, ie 6.3dB for eight wavelengths. The angles (from the run of the antenna wire) at which radiation is at a maximum for different antenna lengths, together with the expected gain, are as below:

| Wavelengths | Angle (°) | Gain over a dipole (dB) |
|---|---|---|
| 1 | 54° | 0.5 |
| 1.5 | 42 | 0.9 |
| 2 | 36 | 1.3 |
| 2.5 | 33 | 1.8 |
| 3 | 30 | 2.2 |
| 4 | 26 | 3.0 |
| 5 | 22 | 4.0 |
| 6 | 20 | 4.8 |
| 7 | 19 | 5.5 |
| 8 | 18 | 6.3 |
| 9 | 17 | 6.9 |
| 10 | 16 | 7.5 |

In practice it will be found that the radiation from the end of the antenna which is furthest from the feedpoint is greater than that from the feed end. This is because the radiation lobes towards the far end are due to the forward-going wave along the wire, whereas from the feed end the radiation can only be due to the wave that is reflected from the far end. Loss by radiation together with any ohmic losses will make the returning and reflected wave considerably weaker than the forward-going one. It is therefore best to 'point' long-wire antennas towards the preferred direction of radiation or reception.

Many years ago the author witnessed a practical demonstration of long-wire power attenuation. L H ('Tommy') Thomas, G6QB, (who is sadly a 'silent key') put up a really long wire antenna which ran around his local golf course. This wire was at least 2000ft (610m) long, and most of it was only 20ft (6m) above the ground. By using a neon bulb tied to the end of a long bamboo pole it was possible to go along beneath this monster and locate the points of RF voltage maxima (ends of half-waves). About halfway along the wire length the neon failed to strike at all, and right at the end of the wire the RF was so feeble that the wire could be safely touched. Grounding the antenna at the end of such a long run would not affect the performance, so end insulators would not be needed!

## Using counterpoise wires

Simple end-fed wires have acquired a reputation for inducing RF feedback problems such as microphone 'howl' and RF in mains wiring etc, together with ATU matching difficulties. There will be some radiation from the wire where it enters the house but this may be largely overcome by using counterpoise wires a quarter-wavelength long. Such wires will provide a useful 'earthy' connection at the ATU which will be far superior to the usual earth-wire system.

An upstairs shack can be particularly bedevilled by earthing problems, for often the run of the earth wire, even when made with heavy-gauge wire or flat strip, has a considerable inductance and will be long in terms of wavelength on the higher-frequency bands. Excessive RF in the shack may give rise to 'hot' equipment, a nasty phenomenon where supposedly earthed metal cases can give the operator an unpleasant RF burn when touched during transmission.

The various long wires used by the author over the years were 'tamed' completely when counterpoise wires were connected to the ATU. The arrangement shown in Fig 28(a) is adequate for most situations and it includes an inductance L (with tap points) and a variable capacitor C, either of which may be used in series with the antenna wire to remove unwanted reactance at the feedpoint end of the wire. The jumper (J) will be used on those bands where reactance does not present a problem.

Reactance problems are usually revealed when it seems almost impossible to bring down the SWR between the ATU and the equipment to a sensible figure. It also may show as very 'sharp' tuning of the ATU. A 'sensible' SWR reading means one which is something between unity and 1.5:1.

The counterpoise wires are cut to a quarter-wavelength for each band, and are best made with PVC-covered multi-strand flexible wires. They may be hung down outside the shack window or instead arranged to run inside the house. They can be put along skirting boards, up to the picture rail (a feature absent from so many modern houses) or under the carpets. Such counterpoise wires will have a considerable RF voltage at their ends when the band they are cut for

is in use and, if the output power is in excess of 50W, it is suggested that their ends are bent over and taped.

L A Moxon, G6XN, describes the use of very short counterpoise wires, which can be brought to resonance with series loading coils, in his book *HF Antennas for All Locations*. The present author has tried that system but found it rather critical to set up. Such loaded wires were also found to have a narrower bandwidth than full quarter-wavelengths. Their use is, however, tempting when faced with the problem of arranging for a 66ft (20m) counter-poise wire (for 3.5MHz) to run inconspicuously through the house!

Counterpoise wires cut to the formula length seem to be effective but they can be set up more accurately with a dip oscillator. To do this, one end of the wire is connected to earth and at that end a half loop in the wire is loosely coupled to the DO coil. A receiver tuned to the wanted frequency, or better still a frequency counter coupled to the DO, will be more accurate than the calibration scale of the DO alone. If no dip oscillator is available three wires for each band could be used, two of these being cut either a few inches longer or shorter than the formula length.

## Ribbon-cable counterpoises

A more elegant way to fabricate counterpoises for several wavebands is to use a length of multi-conductor ribbon cable. This cable is obtainable in 10-way format (or even 20 or 30-way!). Such cable uses stranded 14 x 0.013mm tinned copper wires which are conveniently colour coded.

The use of this ribbon as a four-band counterpoise is

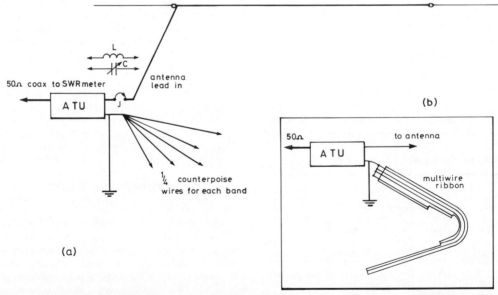

Fig 28. (a) A typical end-fed wire and counterpoise system. The 'jumper' wire is normally connected as shown but, if the antenna displays a capacitive or inductive reactance at certain frequencies, L or C can be inserted in series to cancel it out. (b) How a multi-wire ribbon cable can be cut to provide several quarter-wave counterpoise wires

shown in Fig 28(b). A piece of ribbon is first cut to be a resonant quarter-wave on the lowest frequency band to be used, and it then has sections cut away to make quarter-wavelengths left behind for the remaining bands. Any spare wires can be used to 'broadband' the system by having additional wires that are a little longer or shorter than the calculated midband length.

It is best to splay out the end of each quarter-wave section for about 6in. (15cm) to minimise possible flash-over problems at these points, especially when the transmitter power is high (more than 100W). A counterpoise system made in this manner can be easily hidden and 'lost' beneath carpets etc.

Counterpoises of this type do not contribute towards the radiation efficiency of the antenna, their purpose instead being to reduce or eliminate any RF feedback or matching problems. The 'earthy' ends of all the counterpoise wires must connect right to the earth connection on the ATU. They are very useful 'artificial earths', and the author cured an RF breakthrough problem with an AF filter used for CW reception by connecting the case of the filter to such a quarter-wavelength of wire. The problem only showed on one waveband so only one piece of wire was needed.

## V-beams

When two end-fed long wires are arranged as a horizontal 'V' and fed out of phase, they make a very effective bidirectional beam antenna which will work on several bands (see Fig 29). The feed to the 'V' can be of open-wire tuned line (or 300ohm ribbon used as a tuned line), or instead simply a pair of wires of equal length which come down to the ATU. These wires can be arranged to be spaced at any distance up to about 6ft (2m) on the lower-frequency bands, and there will be no significant radiation from them. The apex angle of the 'V' is arranged so that the radiation lobes from each leg of the antenna reinforce each other within the 'V' and the outside lobes will cancel.

To get this reinforcement, and therefore the maximum gain, there are certain necessary criteria, the most important being the apex angle which depends upon the number of wavelengths contained in each leg of the antenna. The ideal apex angles for different leg lengths are:

| Leg length (wavelengths) | Apex angle (°) |
|---|---|
| 2 | 73 |
| 3 | 58 |
| 4 | 50 |
| 5 | 44 |
| 6 | 40 |
| 7 | 36 |
| 8 | 35 |

It will be seen that the apex angle reduces by a smaller amount as the number of wavelengths increases, and this

means that a compromise angle can be used which will provide useful gain over several bands. A V-beam with five-wavelength legs on 14MHz (about 350ft or 106.6m) will need an apex angle of 44°. This same antenna's legs will each be 7.5 wavelengths long on the 21MHz band and 10 wavelengths long on 28MHz. The optimum apex angles on 21 and 28MHz are 36 and 32° respectively, so a compromise angle of 35° will give an antenna which should work well on three bands. On 14MHz the vertical radiation angles will be raised from the very low angles obtainable when the apex angle is optimum but it will still be at or below 15° above the horizon.

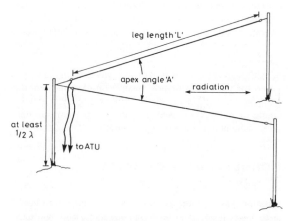

Fig 29. The simple V-beam which is bidirectional. If the apex of the 'V' is not far from the shack, two wires of equal length can be used as a twin-wire feeder. Their spacing is not important

Such an antenna can also be used on the 7 and 3.5MHz bands, although on these bands the gain is reduced and the vertical radiation angles will be higher. This is because the leg lengths are shorter in terms of wavelength and the apex angle is too small.

### Gain and performance

The theoretical gain of a simple V-antenna which uses the correct apex angle is 3dB greater than the gain of a single wire as long as one leg of the 'V'. This means a gain of 7dB for a five-wavelength V-beam. However, in practice the gain realised can be greater than this figure, because it is modified by the mutual impedance between the wires which make up the 'V', and is as much as an additional 1dB with a five-wavelength antenna. At eight wavelengths per leg this additional gain will be almost 2dB, making a total gain for such an antenna as much as 11dB.

Gain of this magnitude is very difficult to achieve with multi-element Yagi beams and represents a power gain of more than 10 times. Even more gain can be achieved by the stacking of two identical V-beams, one above the other, or instead by using two which are broadsided to form a 'W'. However, such complexities put these varieties outside the scope of this book and are of more interest to commercial users.

The three supports which hold up the wires of a V-beam must be at least half a wavelength high at the lowest operating frequency. However, if such an antenna with 35 to 40ft (11 to 12m) supports (about a half-wavelength on 14MHz) is used on 7 and 3.5MHz, its performance on these bands will be similar to that of any horizontal antenna which is relatively close to the ground, and there will be mostly radiation at high angles.

A pair of V-beams, each with 300ft (91.4m) legs and held up by a total of only four 40ft (12m) masts, and using a common three-wire feedline, were designed and used by the author for National Field Day in the early 'fifties. Some care was taken to align the wires using a prismatic compass and the results were spectacular. At that time only QRP operation was allowed in NFD but on 14MHz many Australian and New Zealand stations were contacted with good reports.

Although leg lengths in terms of wavelength have been given for the determination of apex angles and antenna gain, the wires can be of any convenient length just as when single long wires are used. It is only important that both legs of the antenna are of equal length.

## Non-resonant long wires

The simple end-fed long wire is a resonant device and it has standing waves along its length when in operation but, if such a wire is correctly terminated at its far end by the use of a suitable and non-inductive resistor, it becomes non-resonant and additionally unidirectional.

A single horizontal wire can be likened to one half of a two-wire transmission line when the other wire has been replaced by the ground. The characteristic impedance of such a 'single-wire' transmission line, when using normal wire diameters and at a height of between 20 or 30ft (6 or 9m), will lie between 500 and 600ohms.

The semi-unidirectional characteristic of a very long single-wire antenna was discussed earlier in this chapter (see 'Practical end-fed wires'), and it was shown that the radiation away from the far end of the wire resulted in a smaller proportion of the radiation being reflected back towards the feedpoint.

If a resistor with a value equal to the characteristic impedance of the wire is fitted between the far end of the wire and ground, there will be little or no reflected wave. There will be a travelling wave along the wire but no standing waves, and the antenna will be similar to a correctly terminated transmission line but with one important difference – because of the very wide spacing between its conductors (the wire and the ground) it will radiate much of the energy applied at the feedpoint. Approximately half of the power will be dissipated by the terminating resistor but this is not important for the radiation is only needed in one direction.

There will be the usual reciprocity and such an antenna will have the same characteristics for both transmitting and receiving. The gain of a non-resonant wire is similar to that of a resonant wire of the same length so it will not be useful unless it is at least two or more wavelengths long.

The radiation pattern in the horizontal plane is similar to that of a resonant long wire but is modified by the unidirectional characteristic. Fig 30(b) shows the radiation pattern for a two-wavelength, non-resonant, terminated wire antenna. The angles of maximum radiation relative to the direction of the wire are almost the same as the angles

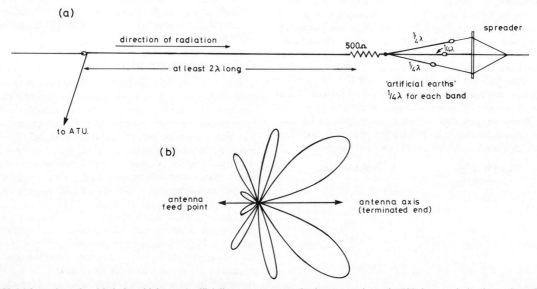

**Fig 30. (a)** A terminated end-fed wire which uses 'artificial' quarter-wave earth wires to terminate the 500ohm non-inductive resistor. This antenna is shown for three-band use. **(b)** The horizontal radiation pattern of a typical terminated long-wire antenna showing its two main lobes

for resonant wires. The antenna feedpoint will show an impedance of about 500ohms and this can be easily matched to a 50ohm equipment impedance by using a simple ATU of the L- or pi-section type.

## A practical design

A most obvious problem when making an antenna of this type is the termination resistor's connection to earth. If the far end of the antenna was dropped to a resistor at ground level, there would then be about 30 or 40ft (9 or 12m) of vertical wire which would radiate in all directions and so ruin the unidirectional property of a non-resonant wire. If instead the resistor was left at the top and a wire was then taken down to ground, the wire would most certainly not be 'earthy' at its top end. The resistor would then present a reactive load to the antenna.

One way to overcome this problem is shown in Fig 30(a) where several quarter-wavelength wires are arranged to behave as 'artificial earths' for each frequency to be used. Only three such 'earths' are actually shown but these may be increased in number. Fig 30(a) shows the 500ohm resistor as an actual part of the antenna top with no indication as to how this may be arranged mechanically. The weight and tension of a long wire would damage any resistor fixed in the position shown in the diagram, so two

alternatives are given in Fig 31. In (a) a rectangular insulating block is used both as antenna insulator and support for the resistor or resistors, and in (b) the insulating block with its resistors hangs beneath and is tied to a ribbed insulator.

## The resistor

The terminating resistor must be able to dissipate just under a half of the transmitting power which, when a basic 100W output transceiver is used, will mean about 45W key-down. Fortunately the duty cycle in the CW mode is only about half of this so a 25W resistor will suffice. The SSB mode has a duty cycle even smaller than that of a CW transmission, so a resistor of about 18 to 20W dissipation will do if only SSB is used.

High-wattage, non-inductive resistors are hard to come by, so instead of such an item a suitable substitute can be built up from several low-wattage components used in a series/parallel combination. The arrangement shown in Fig 31(a) uses 16 carbon-film Hystab 2W resistors. The combination of 470ohm and 560ohm resistors will give an overall resistance of 510ohms and the power dissipation will be 32W. This is more than enough when using a 100W transmitter, either for CW or SSB operation.

Specially manufactured wirewound resistors which are

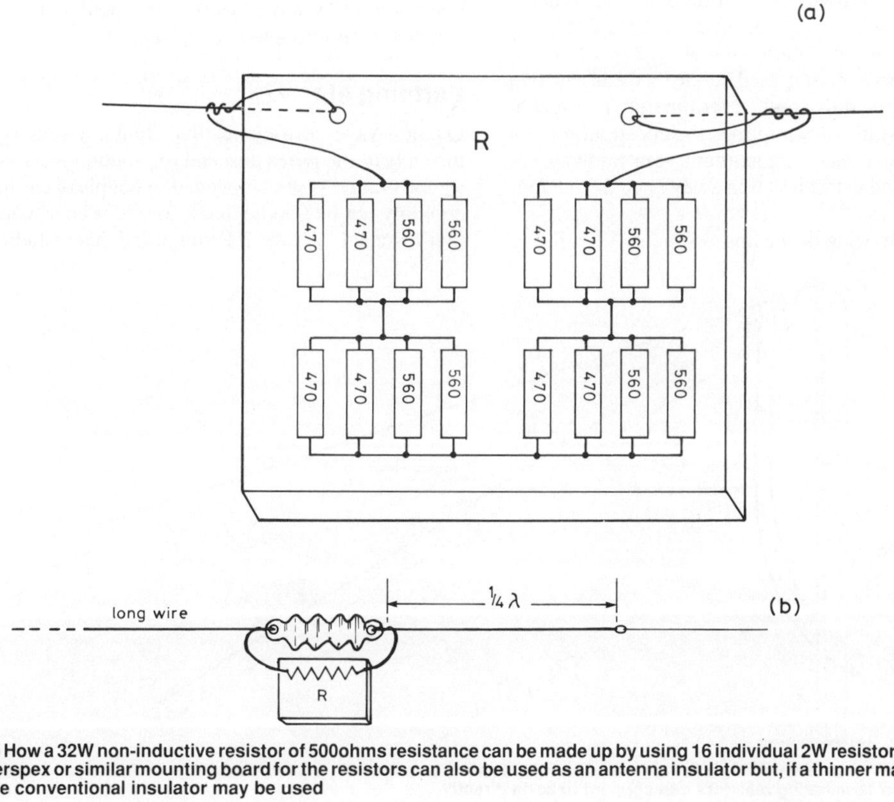

Fig 31. (a) How a 32W non-inductive resistor of 500ohms resistance can be made up by using 16 individual 2W resistors in series-parallel. (b) The Perspex or similar mounting board for the resistors can also be used as an antenna insulator but, if a thinner material is employed, a separate conventional insulator may be used

claimed to be non-inductive are sometimes available on the surplus market, as are also the large carbon types with power ratings of 50W or more. Whichever kind of resistor is used, it *must* be a non-inductive type.

Protection against the weather is also essential when using resistors in an outside environment. The author has used a bank of low-wattage resistors in the manner shown in Fig 31(a) for many months at a location close to the sea, and has found that a liberal coating of the resistors and their connecting wires with clear silicone-rubber sealant has been very effective.

Another writer claims that before it stabilizes this substance contains acetic acid which will attack metals, but its use for the weatherproofing of antenna connections etc over very long periods (years) has never revealed any detrimental effects at the author's location. The heat dissipated by the resistors does not damage or seem to change the physical properties of the silicone rubber once it has stabilized and set. There are of course many other ways to weatherproof resistors etc, and the reader might like to house the resistor in a waterproof plastic container or bind with special tapes such as Sylglas.

## A non-resonant 'V'

The previously described V-antenna, made with a pair of equally long resonant wires, is bidirectional. A V-beam can also be made with two terminated non-resonant wires and such a beam will be unidirectional.

If only one support mast is used (at the feed end) the two wires may be sloped down to their terminating resistors at ground level. This greatly simplifies the construction of a V-beam, and it will have a maximum gain midway between the wires and in the direction away from the feedpoint.

A simplified drawing of the non-resonant 'V' is in Fig 32, where the leg lengths L must be a minimum of one wavelength long at its lowest operating frequency, and the mast height H should be between a half and three-quarters of the leg length. The apex angles for the different leg lengths are similar to those which are optimum for the resonant V-beams. Each terminating resistor has a value of 500ohms but, because the power is distributed equally between both wires, the power ratings of the resistors can be halved from what would be necessary when using a single terminated wire.

Being non-resonant, this V-antenna can be fed with a non-resonant feeder which should have a characteristic impedance of 500 to 600ohms. Such a feeder is easily made from 18SWG wires which are spaced at 3in. (7.5 cm), but any open-wire feeder or even 300ohm ribbon can be used instead as a tuned line. Another alternative would be to use a 9:1 step-up balun at the antenna apex which could then be connected to a 52ohm coaxial feeder. There would not be a perfect match, but the SWR would not be high and the losses would be insignificant.

This antenna will work over a frequency ratio of about 3:1 and give several decibels of gain across this range. Leg lengths of 100ft or 30.5m (1.5 wavelengths on 14MHz), a support height of 60ft (18m) and an apex angle of 80° will make a useful point-to-point antenna for long-distance work on the 14, 21 and 28MHz bands. These dimensions can be reduced to half-size (using the same apex angle) for operation on 28 and 50MHz, and would make a compact antenna to fit into average-sized gardens.

### Earthing etc

Quarter-wave 'artificial earths' are not necessary because the ends of the wires descend to ground level, but instead a good ground system is needed. A couple of earth rods will certainly not be good enough for the proper working of a non-resonant 'V', and it is suggested that in addition there

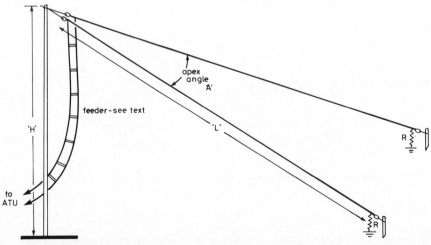

**Fig 32. A terminated V-beam antenna. This is very useful for reliable point-to-point long-distance working. As the antenna wires tilt down to ground level their terminating resistors can connect to earth directly**

should be a minimum of six buried radial wires running back towards the support mast from each of the earthing rods. These radials should be as long as is practicable: say, 35 to 40ft (11 to 12m) each. The resistors can be arranged as suggested in Fig 31, and the smaller 1W types used instead of the 2W 23 by 8mm ones specified for the non-resonant single wire.

Instead of a support mast a tie point on a building (such as a chimney stack) can be used with little loss of gain, for the directivity of the antenna is away from the support point. If the garden runs away from the house to the northwest such an antenna would make an ideal temporary radiator for UK entrants to the ARRL Contest weekends! The fact that non-resonant V-antennas are always grounded via their terminating resistors will be a factor in reducing or even eliminating the build-up of static on the wires. This static on 'normal' long-wire antennas can produce a high noise background in certain weather conditions. A grounded antenna system is also inherently safer than other types.

## The W3EDP antenna

Just who the American amateur W3EDP was the author does not know, for he was not listed in his oldest *International Call Book*, the 1951 edition. What is certain, however, is that before the second world war he devised a simple multiband antenna which used an 85ft (25.9m) wire together with a 17ft (5.1m) counterpoise which was only connected on some bands. The 85ft length may be bent to suit difficult urban locations, and the antenna appears to radiate equally well in most directions.

Such an arrangement was the author's first antenna when licensed in 1946, and it was taken from the pages of an early edition of the RSGB *Amateur Radio Handbook*. When only using a power input of 25W to the 6L6 valve transmitter, the world was worked on 3.5, 7 and 14MHz with the author's version of the W3EDP, which snaked about just above the flat roof of his sea-front QTH. There was no television at that time as the author lived outside the reception area of the transmissions from Alexandra Palace, so the EMC properties of the W3EDP remain a mystery!

The W3EDP antenna, when examined (Fig 33), looks something like an end-fed Zepp with a 68ft (20.7m) top and a 17ft (5.2m) feedline. The feeder is rather unusual in that the wires which make it up need not be parallel. The counterpoise (half the feeder line) can be run outside in any direction, or actually be located in the house, either tacked to a wall or running along the floor! Despite this, it will still have equal and opposite RF voltages and currents to those on the bottom 17ft of the antenna wire, and therefore there will be some cancellation of radiation at the lower end of the 85ft wire. This will help to reduce the RF present in the shack on most bands in a manner which is similar to the function of the quarter-wave counterpoises described earlier in this chapter.

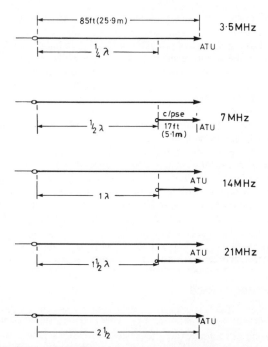

Fig 33. The 85ft W3EDP single-wire antenna will behave rather like an end-fed Zepp on 7, 14 and 21MHz. The counterpoise wire, although not positioned close to the antenna, will nevertheless act like one wire of an open-wire Zepp feed system

## Five-band working

On 3.5MHz the counterpoise is not connected (it would normally join to the ATU earth terminal with a 'croc' clip), and the antenna behaves as an end-fed wire some 19ft (5.8m) longer than an electrical quarter-wave. An advantage here is that the point of maximum radiation (the high-current point) is not in the shack but outside, almost 20ft (6m) away and (hopefully) in the clear. This additional length from the basic quarter-wave also ensures that the impedance at the lower feedpoint end is not too low, so there will be easier matching to the ATU with some inductive reactance which can easily be tuned out.

The counterpoise is connected for operation on 7MHz and the antenna becomes a half-wave (actually it will be about 18in. or 46cm too long) with a widely spaced 'feeder' which is about an eighth-wavelength long. At the feedpoint the impedance will not be very high, nor will it show the very low impedance (36ohms approx) that is present a quarter of a wavelength from the end of a half-wave top. Instead there will be a medium impedance which can be matched without difficulty.

On 14MHz the 68ft which lies beyond the initial 17ft of the antenna wire is within a few inches of being a resonant full wavelength for this band, and the 17ft of 'feeder' is 6in. (15cm) longer than an electrical quarter-wavelength. The impedance at the ATU end will be low, but not so low as it might be were the 'feeder' to be an exact quarter-wavelength.

One and a half wavelengths on 21MHz is almost exactly the same length (actually 67ft 10in. or 20.68m) as a full wave on 14MHz. The high impedance at the shack end of this 1½-wave section (measuring from the far end of the 85ft wire) connects to the final 17ft, which then becomes a part of the feeder which is about ³/₈-wavelength long on 21MHz. This length presents a medium impedance at the feedpoint and is easy to match.

The counterpoise is not required when the antenna is used on 28MHz so the antenna then behaves as a 2½-wavelength end-fed wire. Two and a half wavelengths on this band is exactly 85ft! This band is the only one where there may be matching problems, for the feed end of a 2½-wavelength wire is at quite a high impedance. The introduction of a series capacitor between the antenna and the ATU might help to reduce matching problems for it will electrically 'shorten' the antenna.

## Practicalities

The W3EDP is perhaps one of the simplest and cheapest of all multiband antennas and, as previously mentioned, bending the run of the wire in one or several places to allow its erection in awkward locations need not prevent it from working effectively. This type of antenna is a favourite with some of the QRP fraternity, and with just a few watts easy communication can be achieved all over the UK and into Europe on the two lower-frequency bands.

The antenna is also ideal for temporary or holiday use for it only needs two lengths of wire, the stranded PVC-insulated varieties being the most suitable. No special earthing arrangements need be made, and the 85ft wire can be slung over a roof ridge, a high wall or similar. A small ball of Plasticene fixed to the end of this wire will facilitate such activities!

## Antenna lead-in problems

Arranging the entry of low-impedance twin wire or coaxial feeders into a shack is not very difficult, for such feeders will not normally have standing waves along their lengths. They can be positioned very close to 'earthy' objects and show little or no loss as a result. A simple hole through brickwork or the wood surround to a window will suffice to bring in such feeders, and they may also meander around the house after entry without loss before they reach the radio equipment. However, when end-fed wires or tuned lines are used such simple entry methods are inefficient and lossy, and a much more sophisticated approach to the problem is needed.

The simplest effective way to bring antenna or tuned-feeder wires in, remembering that these wires may carry high RF voltages, is to drill suitably sized holes through the wooden window surround where convenient, and then insert insulating liner tubes. These will prevent the wires

(normally single-strand and uninsulated) from being in contact with the woodwork.

Suitable insert tubes can be made from old ballpoint pens etc, but the author must confess that over the last 20 years his 16SWG copper antenna wire has entered the shack through an uninsulated hole through the wooden window frame. The actual losses which may arise through what may be described by some as 'sloppy' practice are anyway minimal and cannot be measured.

There is, however, the question of the dryness of the wood through which the wire passes. The window used by the author faces away from the prevailing southwesterly rain-bearing winds and is also in a sheltered corner of the property. Additionally there is a large roof overhang, and it is only on rare occasions that any rain drives towards the window in question. Dry wood is quite a reasonable insulator, even at RF, but damp wood can be disastrous! If a 6in. (15cm) length of really *dry* pine has wire 'connections' fixed to each of its ends, it will be found that when shunted across high-power tuned circuits there will be no obvious damping of the circuit and virtually no effect upon it.

Beware of damp wood, however. An amateur the author once visited ran a high-power AM transmitter on 14MHz and, each time that he switched to transmit, smoke came out from the hole where his antenna wire ran through the window frame! Despite this regular pyrotechnic display the amateur in question put out very strong signals to DX, and to the author's knowledge the services of the Fire Dept were never needed!

Many windows in older properties have small panes of glass which are held in place by iron frames, and these windows present considerable difficulties. Access holes cannot be used through the metal surrounds when wires and feeders with standing waves on them are used. Any amateur facing this problem should consider the removal of one of the small panes of glass and the replacement of it with a piece of transparent plastic material such as Perspex. This can be drilled for wire entry before it is fixed to the window frame.

It is possible to remove a pane of glass, drill it by using a special technique (using a paraffin lubricant) and then replace it, but such a procedure requires a considerable patience and skill and cannot be recommended.

A seldom-used but very effective lead-in system for single wires or tuned feeder pairs is shown in Fig 34. Here the window glass becomes the dielectric of a capacitor which is always in series with the wires, and no drilling is needed. This series capacitor will have little effect upon the characteristics of an end-fed antenna or upon lengths of tuned feeder; it will simply make them electrically shorter.

Modern adhesives, particularly the so-called 'super glues', can be used to stick rectangular pieces of thin copper or brass such as 'shim' or foil to each side of the window glass. Each piece must have a small lug to which the wires can be soldered. To reduce any strain upon the 'capacitor', a thin flexible conductor wire should be used to connect with the antenna which will normally be under some tension (see Fig 34). This connecting wire, like all wires going to a window, must have a downward bend on it which will make a run-off point for rainwater and prevent it reaching the metal capacitor plate.

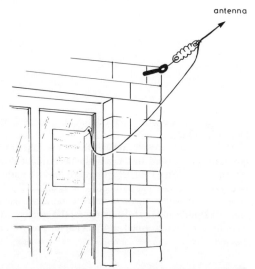

Fig 34. When it is impossible to arrange for a lead-through hole at a window, a capacitor made from two thin copper sheets will be very effective. Some way to earth the actual antenna must be devised for obvious safety reasons

To test this system, a temporary feedthrough capacitor can be held in place by using Sylglas tape or similar. The dielectric constant of window glass is high (between 7.6 and 8.0), so quite a small plate area will provide a useful capacitance. Normal 3mm thick window glass, when used with two plates each having an area of 100cm$^2$, will make a 200pF capacitor. A more useful size for all-band use would be one with an area of 400cm$^2$, which will provide an 800pF capacitor, large enough to have virtually no effect upon the operation of any antenna or feeder on the amateur bands.

This lead-in technique has the disadvantage that the outside wires cannot be earthed from inside the shack and, when long unterminated wires are being used, some way to gain protection from the build-up of static (or worse!) must be devised.

A way to ensure some measure of safety, and also get rid of static charges, is to make a simple spark gap which will connect across the lead-in end of the antenna or feeder and a stout conductor wire which runs down to earth. A string of 10 2W resistors of the carbon-film type, which will make a total series resistance of 200kohms, can be arranged in parallel with such a spark gap.

This resistance between the antenna and earth will in no way affect the antenna performance but will certainly allow the discharge of any static which has built up on the wire. A convenient way to arrange this is illustrated in Fig 35.

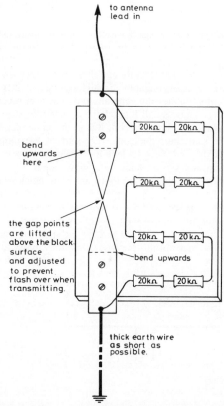

Fig 35. A suitable combined spark and slow-leakage device for use with the antenna lead-in arrangement shown in Fig 34. The eight series-connected 2W resistors will have no effect upon antenna performance but will provide a leak path for small static charges. There is also a conventional spark gap which must be set to prevent flash-over when transmitting

# Chapter 4

# Transmitting loops

Closed-circuit antennas, a variety which are more usually called 'loops', fall into two basic types: those which contain a total conductor length which is small when compared to a wavelength, and those which use a wavelength or more of conductor. Small loops can be likened to large coils and they are generally only used for receiving purposes. These small loops have a current distribution similar to that found in a coil: it is in the same phase and has the same amplitude in every part of the loop. To achieve this end, the total length of the conductor in such a loop must not exceed about 0.1 wavelength.

Very small transmitting loops are both inefficient and difficult to feed. They are very high $Q$ devices and will only operate properly over a very narrow bandwidth. Loops which contain a half-wavelength of conductor (the 'Levy' loop being an example) are perhaps the smallest practical loops for transmitting purposes, but they have a poor performance (1dB down on a dipole at best) and present feed difficulties.

A closed half-wave loop has a feed impedance of a few thousand ohms, and this can only be brought down to 50ohms by making a break halfway around its length. Then of course it is no longer a true loop! It is only when a loop contains a wavelength or more of conductor that it becomes a simple and useful antenna for the amateur. All the examples to be described in this chapter are within this category.

## Folded dipoles

The folded dipole must be attributed to Kraus, W8JK, and it was developed and in use by amateurs in the late 'thirties, becoming very popular in the 'forties. Reference to Fig 36 may help to explain how folded dipoles work.

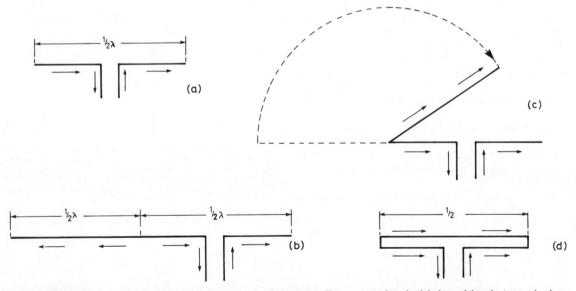

Fig 36. (a) The instantaneous current flow in a half-wave centre-fed antenna. The currents along both halves of the wire top are in phase. (b) The current in the added half wavelength is in opposition to that in the dipole. (c) If this added half wavelength is now brought across as shown, its RF current will no longer be 180 degrees out of phase with the current in the dipole. (d) By joining the end of the added section to the right hand end of the dipole the current in both wires of the new folded dipole will be in phase; each wire having half of the total current.

Fig 36(a) shows a conventional half-wave dipole, the arrows indicating the instantaneous currents along its two wires. They are in phase and there is no cancellation of energy. When another conductor is connected, as shown in Fig 36(b), the current in this new conductor will not flow in the same direction as that in the original dipole. This is because of the rule for reversal of direction of current in alternate half-wave sections along a wire. The fact that the extension to the ends of the dipole is 'folded' back makes the currents in the new section flow in the same direction as those in the original dipole. The currents along both the half-waves are therefore in phase and the antenna will radiate with the same radiation patterns etc as a simple half-wave dipole.

## Feed impedance

The power supplied to a folded dipole is evenly shared between the two conductors which make up the antenna, so therefore the RF current ($I$) in each conductor is reduced to $I/2$. This is a half of the current value (assuming that the same power is applied) at the centre of the common half-wave dipole, so the impedance is raised. The power in watts is equal to $I^2R$ so, by halving the current at the feedpoint yet still maintaining the same power level, the impedance at that point will be four times greater.

| | |
|---|---|
| Dipole: 280W | Folded dipole: 280W |
| Power = $I^2R$ | $I^2R$ (when $I$ = 1A) |
| ($I$ = 2A, $R$ = 70ohms) | $I \times R$ = 280W |
| 4 x 70 = 280W | so $R$ = 280ohms |

This means that a two-conductor folded dipole will have a

feed impedance of 280ohms, which is close to the impedance of 300ohm twin feeder. It can therefore be satisfactorily matched and fed with this feeder, and have a low SWR along the feedline.

If a third conductor is added to the folded dipole (Fig 37), the antenna current will be evenly split three ways and the impedance at the feedpoint will be nine times greater than the nominal 70ohm impedance of a simple dipole. Such a three-wire dipole with its feed impedance of 630ohms will make a good match to a 600ohm feeder. This feeder may be made from 18SWG wires which are spaced at 3in. (75mm).

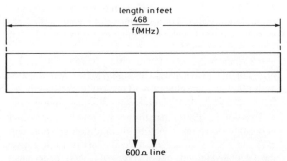

Fig 37. A three-wire folded dipole. If each wire is of equal diameter the total current will then be shared equally between the three wires. The impedance at the feedpoint will then be nine times that of a conventional half-wave dipole (9 x 75ohms = 675ohms) and will be a close match to a 600ohm impedance feedline

A four-wire folded dipole will have a feed impedance of 1120ohms, which is 16 times the impedance of a simple dipole antenna. The feed impedances of the folded dipoles

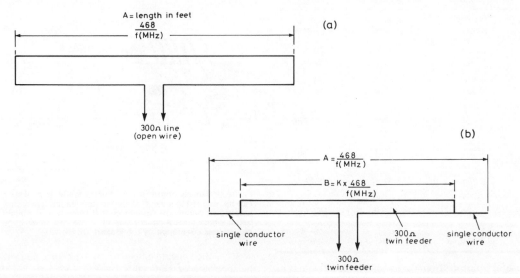

Fig 38. (a) The dimensions of a folded-dipole antenna using wires spaced between 6 and 12in. apart. (b) When 300ohm ribbon is used to make a folded dipole the velocity factor of this material must be taken into consideration when calculating its length. Small end wires are used to make up the full half-wavelength of the folded-dipole top. $K$ is the velocity factor of the ribbon cable and can range from 0.8 (the old type of ribbon) to 0.87 (the newer slotted variety)

so far considered will only apply when their conductor wires are of equal diameter, and are in the same plane. A wide range of step-up ratios may be achieved when tubing elements of differing diameters and spacings are used, but the calculations involved in respect of such arrangements are outside the scope of this book. Such details may be found in *The ARRL Antenna Handbook*.

## Resonant length

A folded dipole which is made with two wires spaced several inches apart will have a resonant length which is equal to that of a simple half-wave dipole (Fig 38(a)), ie $468/f$(MHz) feet. However, if the antenna top is fashioned from 300ohm ribbon or a similar solid dielectric line, the velocity factor of such a line must be taken into account. The older unslotted 300ohm ribbon has a velocity factor $K$ of 0.8 whereas the new slotted variety has a $K$ of 0.87. The resonant half-wave length of $468/f$(MHz) feet must be multiplied by the value of $K$ to determine the length of a folded dipole when it is made with 300ohm ribbon.

This will be shorter than the normal dipole length and it is brought up to resonance by the addition of short wire extensions at the dipole ends (see Fig 38(b)). Table 7 gives the calculated lengths of both types of ribbon feeder when they are used as the main sections of folded dipoles on eight of the amateur HF bands.

**Table 7. Folded dipoles made with 300ohm ribbon**

| Band (MHz) | 'B' Top section using 300ohm slotted ribbon. $K = 0.87$ | 'B' Top section using 300ohm flat ribbon. $K = 0.8$ | 'A' Full length of the dipole = $468/f$(MHz) feet |
|---|---|---|---|
| 3.5 | 113' 1"<br>34.44m | 104' 0"<br>31.69m | 130' 0"<br>39.62m |
| 7 | 58' 2"<br>17.72m | 53' 5"<br>16.3m | 66' 10"<br>20.37m |
| 10.1 | 40' 4"<br>12.28m | 37' 1"<br>11.29m | 46' 4"<br>14.12m |
| 14.15 | 28' 9"<br>8.76m | 26' 5"<br>8.06m | 33' 1"<br>10.07m |
| 18.1 | 22' 6"<br>6.85m | 20' 8"<br>6.30m | 25' 10"<br>7.87m |
| 21.2 | 19' 2"<br>5.85m | 17' 8"<br>5.37m | 22' 1"<br>6.72m |
| 29.94 | 16' 4"<br>4.97m | 15' 0"<br>4.57m | 18' 9"<br>5.71m |
| 29 | 14' 0"<br>4.27m | 12' 11"<br>3.93m | 16' 1"<br>4.91m |

A useful 'spin-off' from the use of a folded dipole is its inherent lower $Q$ and its flatter impedance/frequency characteristic which produces a better bandwidth than that of a simple dipole.

A folded dipole cannot be used at twice its fundamental frequency or at any even multiple of that frequency, as the currents in the two conductors will be out of phase and will cancel. If used on these frequencies it will behave like a continuation of the feedline and the RF currents will be out of phase. On its third and other odd multiples of its resonant frequency, however, a folded dipole will have the proper current distribution and phasing to give effective radiation, and additionally its feed impedance will be close to 300ohms, although the radiation pattern will become that of a centre fed wire three or more half wavelengths long.

## Feeding a folded dipole

The simplest way to feed a folded dipole is with a 300ohm impedance balanced line. This can be either of the solid-dielectric 'ribbon' type or instead made up in the open-wire 'ladder' manner. At the bottom end of this feedline a suitable ATU must be used to match into the low impedance (usually 50ohms) required by most modern equipment.

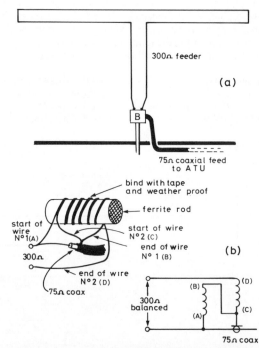

Fig 39. (a) The 300ohm impedance of the feeder to a folded dipole can be reduced to a value of 75ohm (unbalanced) by using a 4:1 step-down balun. (b) Constructional details of a suitable 4:1 balun which uses a short piece of ferrite rod as its core. This design was developed by L A Moxon, G6XN

It is often inconvenient to arrange for a long run of balanced feeder, and in this case the feed method which is illustrated in Fig 39(a) may be adopted. Here the 300ohm line drops vertically from the centre of the antenna to almost ground level, where it connects to a balun B, which must have a step-down ratio of 4:1, providing an unbalanced output impedance of 75ohms. A length of 75ohm coaxial cable can then be led from the balun to the operating position, being buried if necessary.

An ATU must still be used in order to match the 75ohm impedance of the cable to the nominal 50ohms of most equipments. An ATU will also reduce any third-harmonic content from the transmitter which would be radiated by a folded dipole.

Suitable baluns can be purchased and this is a course to be recommended when high-power operation is contemplated, but efficient baluns can be easily made for power levels of 100W or less. Simple 4:1 and other baluns can be readily constructed from short lengths of enamelled copper wire and ferrite rods of the type which are used in MW/LW broadcast receivers.

The constructional details shown in Fig 39(b) are based on those shown on pp51-52 of *HF Antennas for All Locations* by L A Moxon, G6XN. Mr Moxon has experimented at length with homemade baluns, and he has found that common ferrite rod cores are very suitable for use on the HF bands when used with moderate transmitter power levels. The present author has made several examples of the G6XN balun designs and each one has worked perfectly.

## Balun construction

Two 14in. (35cm) lengths of 18SWG enamelled copper wire (which must be new and not taken from an old transformer etc) should be held together with their ends tightly gripped in a vice. A strong pull on the wires and a rub with a piece of rag will get the wires really straight. They may then be taken from the vice, laid together side by side and bound together very tightly with insulating tape over their centre 10in. (25cm).

This wrapped section is wound tightly on to a short piece of ferrite rod and then taped into position. About 3in. (8cm) of ferrite rod is enough for the balun; after winding there will be some of the rod still unused at either end, the actual winding taking up about 2in. (5cm). The beginning and end of each wire must be identified and marked, and then the coil connections can be arranged as shown in Fig 39(b). Such a bifilar coil wound and connected as described will have an impedance ratio of 4:1 over a wide frequency range.

If a valved output transceiver is used, a 75ohm coaxial cable may be connected to it directly. However, when solidstate amplifiers are used, an ATU must be used to match down to the needed 50ohms.

The finished balun may be housed in a weatherproof box, the connection terminals and coaxial cable connector being protected in any one of the ways already described, ie with silicone-rubber sealant or waterproof tape etc.

A balun should never run hot. Any heat represents wasted power and also indicates that there is a serious

mismatch. Baluns should never be used as step-down transformers when the feeder is of the tuned type; only an untuned feeder that is correctly matched to an antenna must be used. If this rule is not observed there can be considerable overheating and much loss of power. The proper use of a balun will result in some small power loss but this should be not more than about 0.1dB. The feeder SWR should be better than 1.35:1 over the frequency range 3.5-28MHz.

Such a 4:1 balun may of course be connected right up at the antenna feedpoint, but there is little to be said for such an arrangement as the feeder coming down would be of heavy coaxial cable and, being unbalanced, it could be detrimental to the working of the folded dipole.

## Folded dipole construction

The best and most mechanically sound way to construct a half-wave folded wire dipole is shown in Fig 40. A single length of wire (preferably single-strand copper) makes up the two dipole elements, and they can be held apart by the use of a few spreaders which are made from a weatherproof insulating material. The RF voltages towards the ends of folded dipoles are not very high so there is little likelihood of large losses from the use of such spreaders.

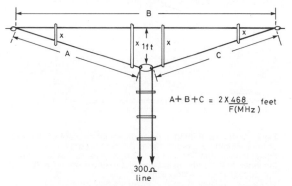

$$A+B+C = 2 \times \frac{468}{F(MHz)} \text{ feet}$$

300Ω line

Fig 40. An 'all-wire' folded dipole. The feeder can be the open-wire type or a length of 300ohm ribbon. A disadvantage of ribbon feeder is that its characteristics can change when it is wet

The fact that the top and the bottom dipole sections are not exactly parallel will have almost no effect upon the performance or the feed impedance of such an antenna. Instead of using end insulators such as those shown in the diagram, lengths of nylon or similar cord can be used as both insulators and supports. A horizontal folded dipole, like a simple half-wave dipole, may be arranged to slope or be vertical, and will then show the same changes in its radiation patterns and polarisation as the dipole. A folded dipole has no power gain advantage over a basic half-wave dipole.

When 300ohm ribbon is used as the whole or just a part of the radiating element (and as the feeder) of a folded dipole, some care is needed to reduce any physical stress upon the ribbon and the connections. Fig 41(a) illustrates how a suitable centre connecting block can be made. A material such as the ubiquitous Perspex can be easily cut or drilled and it also has adequate insulation characteristics for most weather conditions. Dabs of silicone rubber or some similar water-repelling substance should be applied to the soldered connections at the dipole centre.

A folded dipole made from both ribbon and wire sections as already described (also see Table 7) presents mechanical problems where they join, but the use of Perspex strips as shown in Fig 41(b) is suggested to overcome these difficulties. This approach does not place undue strain upon the junction of the end wires with the ribbon, and the friction on the ribbon where it is threaded through the three narrow slots will prevent slippage.

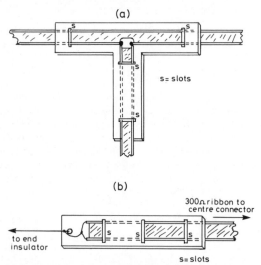

Fig 41. (a) How a 'T' shaped piece of insulating material (Perspex or similar) can be used to make the centre section of a folded dipole made with 300ohm ribbon. (b) A suitable way to connect the inner 300ohm ribbon section of a folded dipole, Fig 38(b), to the single-wire end extensions

## Folded dipoles as beam elements

The folded dipoles described are simple antennas, but they are often used as a part of a complex beam antenna. When parasitic elements are arranged to be in close proximity to a dipole radiator element (ie less than quarter-wavelength spacing), they will affect and bring down the feed impedance at the dipole centre to a low value. A simple half-wave dipole, when being used in a beam antenna which has just a single parasitic reflector spaced at 0.1 wavelength, will have its centre impedance reduced to around 15ohms.

By using a two-wire, folded-dipole radiator instead of the simple dipole, the feed impedance will be raised by four times this figure: up to 60ohms. There will then be a fair match to either 50ohm or 75ohm coaxial cable or twin lead. Many of the commercially made VHF and HF beam antennas use one of a variety of folded-dipole variations to allow ease of matching.

## Full-wave quad loops

The full-wave quad (square) antenna was originally designed and described by Clarence C Moore, W9LZX, in the 'forties. In the 'thirties some 5m band experimenters were using circular full-wave loops as antennas. These were made from copper tube and they are no doubt the ancestors of the quad.

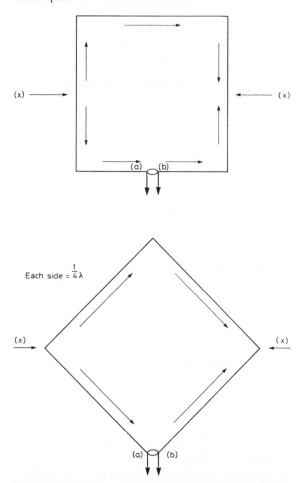

Fig 42. The instantaneous current distribution along the sides of a full-wave quad loop. When the loop is bottom-fed, as is shown in both these examples, its radiation will be horizontally polarised. The points x are at high impedance and voltage which means that the quad arranged as a diamond must have very good insulation at two of its corners x

A quad was derived from a 'pulled out' folded dipole: as such it retains the RF current characteristics of such an antenna (see Fig 42) with in-phase currents along both its top and bottom sections. Each side of a quad is a quarter-wavelength long, and the antenna may be arranged either as a square or as a diamond. The two examples illustrated in Fig 42 are fed at their bases and they will show horizontal polarisation. When fed at a point half-way up either of their sides the polarisation will be vertical. The length of conductor used to achieve resonance in a quad is greater than just two half-wavelengths (936/f(MHz) feet), and instead the expression 1005/f(MHz) feet must be used when designing this type of antenna. Practical conductor lengths for quad antennas for use on the amateur HF bands are given in Table 8. Measurements for a quad on 1.8MHz are not shown, for such an antenna can only be used when supports more than 140ft high are available!

## Quad characteristics

Unlike a folded dipole the feed impedance of a quad is only about 100ohms, and additionally the antenna exhibits some slight gain over a half-wave dipole. The direction of maximum radiation is at right angles to the plane of the quad loop (ie looking through the loop shows the directions of greatest radiation). The gain should be about 1dB, which represents a power gain in two directions of 1.26 times and is a useful feature of the single-loop quad antenna. There are quite deep nulls in the plane of the loop, which are more pronounced than the nulls off the ends of a dipole.

**Table 8. Practical conductor lengths for quad and delta antennas**

| Band (MHz) | Length of delta loop or quad | Length of each side | |
|---|---|---|---|
| | | Quad | Delta |
| 3.6 | 279' 1"<br>85m | 69' 8"<br>21.25m | 93'<br>28.33m |
| 7 | 143' 6"<br>43.76m | 35' 10"<br>10.94m | 47' 10"<br>14.58m |
| 10.1 | 99' 6"<br>30.32m | 24' 10"<br>7.58m | 33' 2"<br>10.1m |
| 14.15 | 71' 0"<br>21.64m | 17' 9"<br>5.41m | 23' 8"<br>7.13m |
| 18.1 | 55' 6"<br>16.92m | 13' 10"<br>4.23m | 18' 6"<br>5.64m |
| 21.2 | 47' 5"<br>14.44m | 11' 10"<br>3.61m | 15' 10"<br>4.81m |
| 24.94 | 41' 2"<br>12.55m | 10' 3"<br>3.13m | 13' 9"<br>4.18m |
| 29 | 34' 9"<br>10.56m | 8' 8"<br>2.64m | 11' 6"<br>3.52m |

A further advantage of the quad loop is that, being a closed loop, it is less susceptible to the effects of the ground than a half-wave dipole. At a height above the ground of a half-wavelength the main radiation lobes of a quad antenna are about 4° lower than those of a half-wave dipole at the same height. At ⅛-wavelength the radiation angle is almost 10° lower. Down at a quarter-wavelength above ground a dipole becomes almost useless, for most of

its radiation will be upwards, but a full-wave quad will still have its main radiation lobes 40° above the horizon. This represents a 'first skip' distance of about 400 miles.

The influence of near objects such as trees or buildings upon the characteristics of quad antennas is small. This means that such antennas can often be used to good effect even when located in house roof spaces etc.

## Practical quad loops

The two basic configurations for quad antennas are given in Fig 43. When arranged as a square, as in Fig 43(a), bamboo or glassfibre spreaders (or 'spiders') will be needed. The high-voltage points along the antenna will be at the centre points of the vertical wire sections, and therefore well away from the ends of the spreaders at A, B, C and D. The impedance at these corner points will not be high; little in the way of insulation will be required where the wire is tied to the spreaders.

If bamboo is used, it should be waterproofed by the application of several coats of polyurethane varnish. Despite the shiny appearance of this wood, it will, if untreated, rapidly absorb moisture and will rot. The ends of any piece of bamboo will also need special weatherproofing treatment. The centre block shown in Fig 43(a) can be made from thick Perspex or ½in. (12mm) marine plywood. If wood is used it must be varnished. A pair of 'U' bolts can be used to hold each spreader in position, and a small rectangle of almost any kind of insulating material (so long as it is weatherproof) will suffice at the feedpoint.

The arrangement in Fig 43(b) will need more space, for its height and width are the diagonals of a square. Furthermore, the RF voltages and impedances will be high at the points B and C. This means that there must be good insulation at these positions. Small ceramic 'stand-off' insulators or similar can be used at A and at the ends of the wooden cross bar. The feedpoint insulation is not critical, as it is at a relatively low impedance, and here even a well-varnished block of hardwood should serve.

A dip oscillator can be used to check a quad for resonance. A single-turn coil across the feedpoint (feeder removed) can be coupled to this piece of test equipment to determine quad tuning.

The quad feed impedance of 100ohms is awkward to match with commonly available feeder. In order to avoid the use of tuned stubs or matching transformers, one simple way to feed such an antenna is by employing two equal lengths of 50ohm coaxial cable (see Fig 43(c)). This arrangement will provide a 100ohm balanced feedline which can be easily matched to the equipment via an ATU. This type of line, as it is made up from coaxial cable, can be safely buried. Of course, open-wire tuned feeder (or 300ohm ribbon used as tuned line) may be used but only with an ATU.

The 1dB gain of a quad makes it a useful and simple bidirectional beam antenna, and it only need be turned through 90° to realise all-world coverage. Its deep nulls are

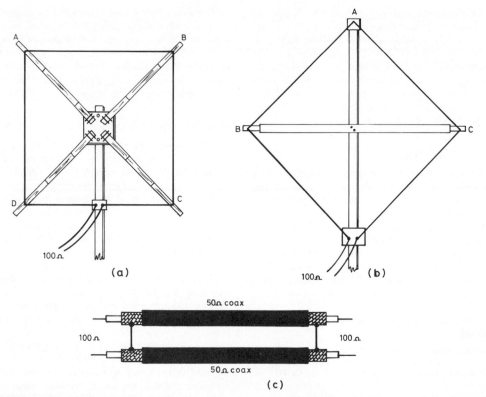

Fig 43. (a) A square quad arrangement using bamboo spreaders. The RF voltages will not be high at the points where the wires are fixed to the bamboos, so no special insulating arrangements are needed. (b) With this arrangement point A is at low impedance but points B and C will require good insulation. (c) Two equal lengths of 50ohm impedance coaxial cable can be connected in this way to provide a balanced 100ohm feeder for a quad antenna

also useful and can be utilised to reduce the strength of unwanted signals.

## Full-wave delta loops

A delta loop is also a closed-loop antenna and it contains a wavelength of conductor wire. It also has several features which are common to the quad antenna. A delta loop is normally arranged in the form of an equilateral triangle, having either one side at the top or at its base. Large delta loops are popular for use on the LF bands for they are much easier to set up mechanically than a large quad antenna. They have a smaller internal area than a quad which reduces their comparative gain.

The greatest enclosed area possible for a full-wave loop would be provided by a circular design, but the mechanics of a circular loop are considerable! If a delta-loop antenna were to be designed in a flattened form with a long base and an obtuse top apex angle, it would closely resemble a folded dipole and show little gain over the latter.

An equal-sided delta-loop antenna will show a gain of about 0.5dB less than a quad (ie 0.5dB over a dipole) and the two sloping sides of the delta will only contribute a

third of the radiated power. The remaining two-thirds come from the horizontal side.

There are five basic arrangements for the orientation and feeding of delta loops. These are shown in Fig 44: types (a), (c) and (d) will give horizontal polarisation and (b) and (e) will give vertical polarisation. The geometry of a delta loop, which may be used with either a horizontal side at its top or its base, determines the effective mean height of the antenna. This effective height is greatest when one of the delta sides is positioned at the top (see Fig 44).

The half-wave ends (ie the points of high voltage and impedance) are indicated in each of the five arrangements shown in Fig 44. The true effective mean height may be calculated as follows.

When one side is at the *top*, the effective height is the height from ground to the highest point of the antenna minus $1/12$-wavelength. When one side is at the *base*, the effective height is the height from ground to the highest point minus a quarter-wavelength.

The total conductor length in a delta loop is the same as an equivalent quad on the same frequency, ie $1005/f(\text{MHz})$ feet or $306.3/f(\text{MHz})$ metres.

Like a quad the maximum radiation from a delta loop is at right angles on both sides of the plane of the loop. Its feed

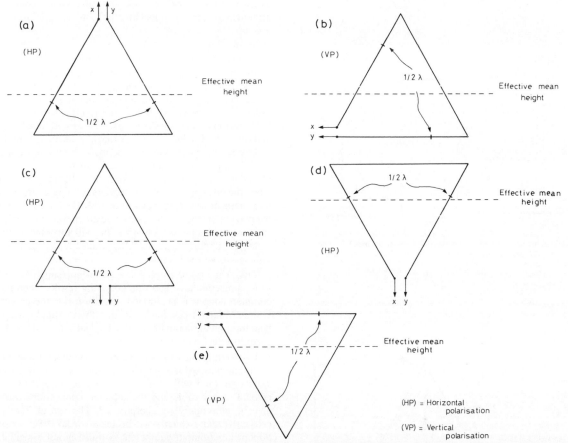

**Fig 44. Five different delta loop configurations. The arrows indicate the high-voltage points at the ends of each half-wave. Versions (d) and (e) provide the best effective height above ground for a given mast height. The two vertically polarised versions shown in (b) and (e) are the best for long-distance work, for they both have their maximum radiation at low angles to the horizon. Types (a) and (c) have high-angle radiation (80°) and would only be suitable for short-distance communication. The delta loop at (d) has a radiation angle of about 47° and will normally have a first-skip range of 400 miles**

impedance will range from about 70ohms in the case of an equilateral delta to more than 100ohms when the antenna is 'flattened' somewhat. Another similarity with a quad is the delta's pair of radiation nulls in the plane of the antenna.

## Performance

The performance of quad and delta-loop antennas is largely determined by their height above ground. Each of these antennas has an 'effective mean height' (see Fig 44 for delta loops), and this equates with the height of a half-wave dipole above ground when it is cut for the same band. In the case of a quad antenna the effective mean height will be at a point halfway up its vertical wires. For long-distance communication a half-wave dipole must be at least a half-wavelength above ground, and this distance will still apply when considering the effective mean heights of loops.

The radiation angles of delta loops are also greatly influenced by their physical arrangement and the positions of their feed points. Only two of the five delta loops illustrated in Fig 44 are really useful for DX working. Type (b) has its maximum radiation at 27° and type (e) has an angle of only 20°. Both of these loop types are vertically polarised.

The horizontally polarised versions (a), (c) and (d), however, have radiation angles to the horizon of 80° and 47° respectively. This means that types (a) and (c) are only useful for very short range communication. Type (d) will have a 'first-skip' range of about 400 miles, whereas the two vertically polarised deltas (b) and (e) will have first-skip ranges of 600 miles and almost 1000 miles.

Unfortunately type (e), which is the best for DX work, needs two supports and it also has an inconveniently positioned feedpoint. Type (b), which only requires a single support and has its feed at one lower corner, is the best configuration for ease of construction when a large delta loop is contemplated for use on one of the LF bands.

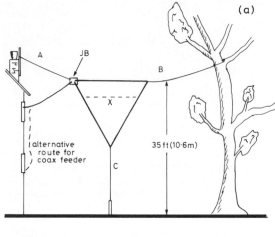

(a)

X = mean effective height / (almost 1/2 λ)

JB = junction block

A
B   } = nylon cord
C

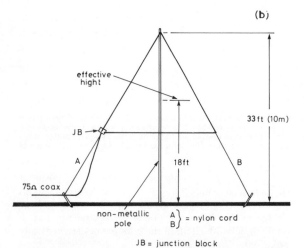

(b)

JB = junction block

Fig 45. (a) A practical delta-loop antenna for 14MHz based upon the type shown in Fig 44(e). It is suspended from two supports. Its radiation angle is only 20°. (b) This version uses a single support mast and is the arrangement shown in Fig 44(b). Its effective height is, however, only 18ft (5.5m) whereas the antenna shown in (a) has an effective height of almost a half-wavelength

## Practical delta loops

Two of the several ways to set up a delta-loop antenna are given in Fig 45, both of these being 14MHz designs. The example shown in (a) has a mean effective height of almost a half-wavelength, and is arranged to provide low-angle radiation. The rather awkward position of the feedpoint is overcome when it is placed not too distant from the house.

The feeder must not drop down vertically when using this arrangement, or it will unbalance the system and detune the antenna.

In Fig 45(b) a single 33ft (10m) support pole is all that is needed, the lower ends of the delta being held in position by nylon cords. In this antenna arrangement the feeder can safely drop down and run along at ground level or be buried. Conventional insulators are not required, as the voltages at the corner angles of delta-loop antennas are not high. Nylon or Terylene cords are fine as both insulators and supports. The junction blocks are also located at points of low RF potential; they can be made from almost any insulating material which will shed moisture.

iA variation of the antenna shown in (b), which can be used when the physical size of the loop and its support are very large at the lower frequencies (7 or 3.5MHz), is one where the two upper sides of the loop come down from the centre support at an angle of about 45°. This will allow the use of a shorter support mast; on 7MHz a mast height of about 50ft (15m) will suffice.

Trees can also be employed as delta-loop supports, and some amateurs have had fine operating results when the complete loop was positioned actually inside the branch system of large trees. Such antennas then virtually disappear through the summer months; the leaf growth does not seem to affect their performance.

Fig 46 illustrates a suitable connection block for delta-loop antennas. Almost any insulating material which is weatherproofed will do for this – the actual end of the coaxial feeder cable and the soldered connections must also be thoroughly weatherproofed. The use of 75ohm coaxial feeder (which must be taken to an ATU when 50ohm input/output equipment is used) is not essential, and, like many other antennas so far described, a tuned feedline can be employed.

The use of delta loops on frequencies far removed from their design frequency is not recommended. They are not multiband antennas. One radio club which is well known to the author put up a big delta loop cut for use on the 3.5MHz band one Field Day, and the operators soon discovered with some dismay that their efforts with this on the higher-frequency bands were unrewarding! Their score rate was so low that, in desperation towards the final hours of the contest, a trapped dipole was pressed into service.

## The grounded half-delta antenna

This antenna was developed by John S Belrose, VE2CV, and described by him in the American magazine *Ham Radio* in May 1982. It has received scant attention in Europe, for a large plot of land and a very high support tower are needed when the antenna is designed for 1.8MHz use as in the original article.

The basic features and dimensions of the half-delta are shown in Fig 47. The most important element in its design

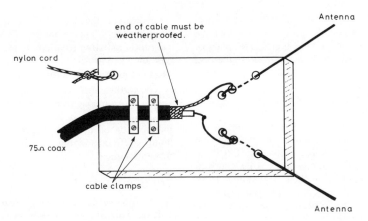

**Fig 46. A suggested connection block for delta-loop antennas**

is the vertical and grounded end support which acts as the main radiator. Belrose used a lattice tower 100ft (30.5m) high, the top of which was connected to a sloping 206ft (62.79m) wire which ran to almost ground level. These dimensions are for 1.8MHz operation, but tower heights of 50ft (15m) and 25ft (7.6m) with correspondingly smaller sloping wire lengths can be used on the 3.5 and 7MHz bands.

The inventor of the grounded half-delta claims that it is one half-section of a full-wave equilateral delta loop, the missing half-wave being the ground reflection. A heavy ground-return wire is needed between the base of the tower and the feedpoint, and in addition there must be really good and effective earth systems at the tower base and also the feedpoint. The vertical section and the sloping wire together add to something more than a half-wavelength, and the feed impedance is about 50ohms. Some 'cut and try' is needed to bring the half-delta into resonance and some British workers have found that the published Belrose dimensions are a little short.

On its fundamental design frequency the grounded half-delta is an efficient low-angle radiator, having a horizontal

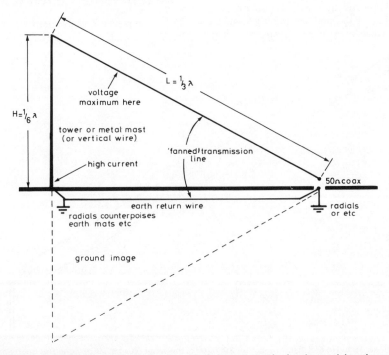

**Fig 47. The VE2CV grounded half-delta antenna. The earth return wire is important as is also the provision of good earth systems at the base of the vertical section and at the feedpoint**

radiation pattern similar to that of a quarter-wave vertical Marconi antenna. However, its pattern shows that the radiation is somewhat greater at right angles to the plane of the antenna, although there are no real nulls as in the case of quad and delta-loop antennas.

An additional feature of the half-delta is its good performance on all the harmonic frequencies of its resonant frequency. On these harmonic frequencies the maximum radiation comes from the ends (in the plane of the sloping wire) as two rather broad lobes, but again there are no really deep nulls in the radiation pattern. On these harmonic frequencies the feed impedance is far removed from 50ohms and can range from 100 to 1000ohms. An L-section matching unit is needed at the feedpoint to overcome this problem, and it will need adjustment for each different band. This complicates its use on the harmonic frequencies and may be one reason why the antenna is not popular.

## Is it really a loop antenna?

The author has tried several experimental versions of the VE2CV half-delta, and as a result has formulated his own theory as to how such an antenna works. Perhaps the simplest explanation is to regard the antenna as a grounded vertical which is fed at the top. The 'feeder' is the sloping wire; this will not contribute very much to the overall radiation. It will work against ground and for much of its length it is close to it in terms of wavelength. At the lower end (the high-current point) the wire is very close to the ground indeed, and little radiation will occur there.

Further along its length the current diminishes, and the impedance and voltage increase to reach a maximum at the

point which is the 'top' of the quarter-wave radiator (the vertical section plus about 25ft (7.6m) of wire). There is very little radiation from points of high RF voltage along any antenna, so this means that almost all the radiation must come from the vertical 100ft tower section.

The earth-return wire enhances the effectiveness of the single-wire 'feeder', and the good ground connection (buried radials etc) at the base of the vertical section also increases the efficiency of the antenna. This explanation is humbly offered and may be quite wrong, but to the author it seems logical.

## Try one yourself

An antenna of this type, when designed for 7MHz, will only need a 25ft metal mast or tower (or even a thick wire coming down from a wooden pole) and about 50ft (15m) of garden length. The suggested length for the sloping wire on this band is 52ft (15.85m). Such an antenna will work on 7, 14 and 28MHz. The author's favourite method of feed, tuned lines, was used by his friend the late G3CMN with his 3.5MHz version, and he obtained quite outstanding results, especially on 3.5 and 21MHz. By using a tuned feeder line, the matching can be done in the shack at the ATU without recourse to the complicated remote switching and tuning of an L-section matcher at the foot of the sloping wire.

## The terminated tilted folded dipole

This antenna seems to have been devised in the late 'forties by Capt G L Countryman, USN, W3HH, and his first

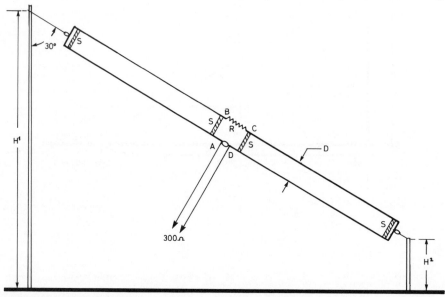

Fig 48. The T2FD (terminated tilted folded dipole) antenna. A version of this antenna which is designed for operation on 7MHz and above will only require a mast height (H1) of about 35ft (10.7m) and a small end support (H2) 6ft high. It is also considerably shorter than a full half-wavelength and will fit into a small garden

article (which showed its potential for amateur use) appeared in the June 1949 issue of the ARRL journal *QST*. Further articles on this topic by the same writer appeared in *CQ Antenna Roundup 1963* on pp68 and 70. The antenna's name is usually abbreviated to 'T2FD'.

It was used at the Long Beach Naval Station, California, with considerable success, its radiation pattern and field-strength measurements there being superior to those obtained from a Marconi antenna. US air bases in the Pacific and several MW broadcast stations in the USA and Japan have also used T2FD antennas. The author's use of T2FDs goes back to about 1951 when a 12ft (3.66m) long 'bedroom' version worked into the USA on 14MHz with ease!

Fig 48 shows the principal features of the T2FD. It bears a superficial resemblance to an ordinary folded dipole, but its dimensions, the use of a non-inductive terminating resistor R and the all-important 20 to 40° tilt result in an aperiodic or non-resonant vertically polarised radiator which has a useful frequency ratio of at least 4:1.

A T2FD designed for 7MHz will work satisfactorily on all frequencies up to 30MHz, to a lesser degree even on half its design frequency (3.5MHz), and can be easily fed with 300ohm impedance untuned line.

When set up at its optimum slope angle of 30°, the T2FD displays an almost-omnidirectional, low-angle radiation pattern similar to that of a vertical quarter-wave Marconi antenna. There is, however, some reduction in field strength at the high end of the antenna, and this must be kept in mind when considering its erection and use.

The antenna is useful in cramped locations, for on its design frequency it is somewhat shorter than an equivalent half-wave antenna. On 7MHz a half-wave is about 66ft (20m) long from end to end but a T2FD will only be about 47ft (14.33m) in length. To be effective, a 7MHz half-wave antenna must also be held up at each end to a height of at least 60ft (18.3m), but the T2FD only needs a single support 36ft (11m) high and an additional short 6ft (1.8m) pole at its low end.

## Design criteria

The author has not yet discovered any literature which really explains just how the T2FD antenna works! Countryman (W3HH) jokingly described the antenna as a "squashed rhombic" but gave no real information as to the theory which led to its design. The TF2D seems to use its terminating resistor to broadband a folded dipole, but just how its dimensions were deduced must remain a mystery.

There are, however, some rigid design parameters to consider when a T2FD is assembled. The length of each leg (when measuring from the centre of the wires across the end spreaders to the feedpoint or the terminating resistor) should be 50,000/$f$(kHz) x 3.28 feet. The total top length and the lengths AB and CD (see Fig 48) will be twice this calculated length. The frequency $f$ is the lowest operating frequency of the antenna although, as has been mentioned,

a T2FD will work with reduced efficiency at half this frequency. The spacing between the two radiator wires D in feet can be found by dividing 3000 by the frequency in kilohertz and multiplying this result by 3.28.

The terminating resistor *must* be non-inductive if a genuine aperiodic antenna which will work over a large frequency range is required. However, the antenna will still work if an inductive (wirewound) resistor is used, but then it becomes resonant on one or more frequencies. In addition the feeder must then be used as a tuned line, the 'flat' 300ohm impedance at the feedpoint being lost.

The terminating resistor value is to some extent determined by the impedance of the feedline used. When using 300ohm twin lead to feed the antenna, the optimum resistor value is about 400ohms, although any resistance value between 375 and 425ohms will work well. With 450ohm open-wire feedline a 500ohm resistor is satisfactory, and the use of 600ohm line requires a 650ohm resistor.

Some experimenters have fed the T2FD antenna with low-impedance line (including coaxial cable), but then the terminating resistor value becomes very critical and must be within 5ohms of optimum. The terminating resistor must dissipate about 35% of the transmitter output power. This may seem to be a serious power loss, but in fact it only represents a signal loss of from 1.5 to 2dB (below half an S-point) which is more than compensated for by the low angle of radiation from the antenna.

## Constructional points

A T2FD antenna can be assembled by using two wires of equal length, each wire making up one of the sections AB and CD (see Fig 48). Their lengths for each band are given in Table 9 where the spacing distances D are also shown. The heights of the antenna supports on different bands have not been calculated, but a few moments spent with a ruler and a sheet of graph paper will reveal these. The mast height of 36ft (11m) for a T2FD designed for the 7MHz band may of course be interpolated to discover the heights needed on other bands.

**Table 9. Terminated tilted folded dipoles**

| Band (MHz) | Length AB and CD (also top length) | Spacing D |
|---|---|---|
| 1.8 | 182' 2" (55.54m) | 5' 5" (1.66m) |
| 3.6 | 91' 1" (27.76m) | 2' 8" (0.83m) |
| 7 | 46' 10" (14.28m) | 1' 5" (0.42m) |
| 10.1 | 32' 6" (9.9m) | 1' 0" (0.3m) |
| 14.15 | 23' 2" (7.06m) | 0' 8" (0.21m) |
| 21.2 | 15' 6" (4.7m) | 0' 5" (0.14m) |
| 29 | 11' 4" (3.44m) | 0' 4" (0.1m) |

Trees or buildings can also be used as end supports, but when buildings are used there will be some additional attenuation of the radiation from that end of the antenna. A

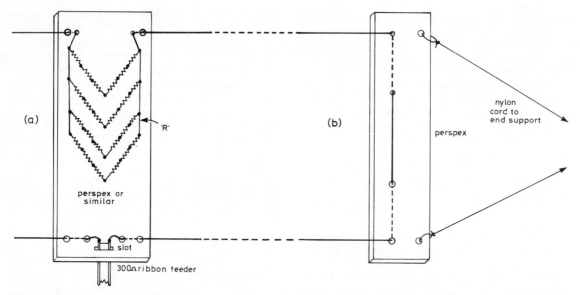

**Fig 49. (a) The centre block of a T2FD, which accommodates a chain of non-inductive resistors, provides a connection point for the 300ohm ribbon feeder and also is an antenna spreader. (b) How the antenna wire is threaded through holes in the two end spreaders. Perspex strips ¼in. (6mm) thick are suitable as spreaders. For a 7MHz version of the T2FD only two end and one centre spreaders are needed**

6ft anchor point at the low end of the antenna can be used for all antenna lengths, its main purpose being to safeguard the wires from the attentions of animals or children.

The author's version of the T2FD, when pulled tight, only needed a centre spreader and a spreader at each end, but there is no reason why additional spreaders cannot be used towards the centre of each leg of the antenna. If the end spreaders are fashioned from a material like Perspex sheet (see Fig 49), the antenna wires can then be threaded through as shown and no additional antenna insulators will be needed.

The antenna centre block (Fig 49(a)) has on it the connections to the 300ohm feeder (the slotted Bofa type for preference) and also the terminating resistor. The latter is made up from a series-parallel arrangement of low-wattage types. The 50% duty cycle of CW (morse) or the even shorter duty cycle of SSB telephony will determine the wattage needed in the resistor chain.

Of course, a continuous carrier is needed for certain other transmissions, such as FSK or FM, and this must be borne in mind when determining the terminating resistor

power dissipation. The resistor will only dissipate about 18W on CW with a power output of 100W, and this will be even less on the SSB mode.

It is suggested that 24 resistors, each with a resistance value of 270ohms and a power rating of 2W, will be more than adequate. The resistors will be well within their ratings under key-down conditions, and on SSB a peak power of 300W can be used.

The resistors are wired into four lines, each of which has six in series (1620ohms per line), and these four lines when paralleled will produce a final resistance value of 405ohms. The resistors must be of the carbon-film type. Weatherproofing (as described in an earlier chapter) is necessary, and the author's last T2FD antenna had its resistors and all the soldered connections liberally coated with transparent silicone-rubber sealant.

Not long after the antenna described was built and in use, the author had an opportunity to purchase several wirewound non-inductive 400ohm resistors, each with a power rating of 80W. If similar resistors (or instead the high-wattage, carbon-compound types) are used, it is

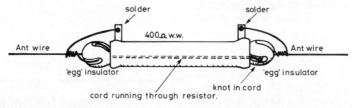

**Fig 50. The way to connect a single high-wattage resistor to the centre of a T2FD antenna. The cord running through the centre of the resistor takes the strain away from the connections. A thermoplastic cord must not be used in this position, and all electrical connections should be waterproofed**

suggested that they are connected as shown in Fig 50.

Two small ceramic or glass 'egg' insulators are positioned at the ends of the resistor and a stout cord (without thermal problems, ie good-quality sash cord) which has been weatherproofed with silicone furniture polish, Damp Start or a similar water repellant can be run between the insulators and through the resistor. This cord will take the full strain of the antenna and prevent any damage to the resistor. The antenna wires also go to the insulators, wrap round a few times and then are bent over to be soldered to the lugs on the resistor. Weatherproofing is only required on the lugs and the soldered connections.

## Feed arrangements

A correctly designed T2FD which has a non-inductive terminating resistor will present a uniform feed impedance right across its frequency range. The antenna described should have a 300ohm feeder which may be taken right to the shack and the station ATU. Alternatively, and especially when the antenna is at a long distance from the operating position, the feeder can connect to a 4:1 balun (as shown in Fig 39) which will bring down the impedance to 75ohms (unbalanced) and allow the use of coaxial cable.

One disadvantage of any aperiodic antenna is that it will radiate any harmonic content of the transmitter output, so it is essential that an additional tuned circuit in the form of an ATU is always used with a T2FD. The T2FD is, however, a balanced type of antenna system and this may assist in limiting EMC problems. Harmonic radiation problems on VHF etc will be small if a low-pass filter is used between the transmitter and the ATU.

## Earthing

The literature on T2FD antennas does not mention the benefits of having a good earth system, but the author and others who have used these antennas have found that they work more effectively when they are positioned over several buried radial wires which connect back to the station ATU. This is not surprising, for the T2FD design seems to have much in common with vertical Marconi antennas which of course rely greatly upon good low-resistance earths.

Chapter 5

# Marconi antennas and ground systems

The Hertzian family of antennas should be at least one half-wavelength above the ground to be effective low-angle radiators (for long-distance communication). This is not difficult to achieve on the higher frequencies, but half-wave dipoles cut for use on either the 3.5 or 1.8MHz amateur bands would need to be up at heights of 130ft (39.6m) or 250ft (76m) respectively to meet this criterion. Not many amateurs in the UK have masts or towers more than 100ft (30m) high, and such a restriction means that to carry out successful DX work on the LF bands a majority must turn to Marconi antenna systems. These systems have a long and distinguished pedigree going back to the start of the 20th century.

"...The invention of the $1/4$-wavelength earthed aerial, where the earth is one plate of the condenser [capacitor], is considered to be the most important of Marconi's early contributions to radio engineering. Since any voltage node is 'earthy' in potential, it may in fact be earthed without affecting the voltage and current distribution..." (*Handbook of Wireless Telegraphy*, Vol II, Admiralty, 1938).

Although apparently simple in concept, the Marconi antenna presents considerable design problems if it is to be an *efficient* radiator. These problems involve the physical length of the antenna (when it is shorter than an electrical quarter-wavelength), and also the effectiveness of the ground conductivity in its vicinity.

For example, a 'short' vertical antenna used on 1.8MHz, such as a 14ft (4.2m), 0.03 wavelength, bottom-loaded wire or rod used over a poor ground system having an earth resistance of about 100ohms, will have a radiating efficiency of only 0.26%! This means that a power of 8W into such an antenna will result in only 20mW being actually radiated. The other 7.98W will be dissipated as heat in the ground resistance and in the resistance of its loading coil.

Fig 51(a) represents a simplified and 'ideal' quarter-wave vertical Marconi antenna. The ground is shown to be a perfect conducting medium, a condition which can only be realised when it is replaced by a sheet of metal which has dimensions that are large relative to the length of the antenna (or less effectively by a large area of salt water).

The ground, if it is a perfect conductor, will behave like

an electrostatic shield and provide an 'image' antenna a quarter-wave below the radiator. This image completes the missing half of a Hertzian half-wave antenna, and earth-return currents will be induced in the ground.

The impedance at the point where a resonant quarter-wave vertical conductor meets the ground is about 36ohms; just a half of the feed impedance at the centre of a resonant half-wave dipole. The current along the quarter-wave vertical antenna is at its maximum at its base and therefore the greatest radiation will take place at this point (see Fig 51(b)). The radiation will be vertically polarised and in the example illustrated will have equal field-strength levels in all directions.

Much of its radiation will be at low angles to the horizon when above a good ground, and this makes the vertical Marconi antenna very attractive for both short-distance (ground wave) and long-distance communication on the lower-frequency bands. The polarisation of an antenna when used for long-distance work does not matter, for the effects of refraction in the ionosphere etc will inevitably induce changes in polarisation.

An interesting observation is that a quarter-wave vertical antenna may be compared with a closed organ pipe. Such a pipe when blown gives a musical note which has a wavelength approximately four times the length of the pipe!

## Practical arrangements

It is seldom possible or convenient to erect a full-sized quarter-wave vertical for the lower-frequency bands, although such antennas are often used on the higher frequencies where they are called 'ground-plane' antennas.

In Fig 52(a) the quarter-wave is in the vertical plane and is shown to be bottom fed (impedance 36ohms). Figs 52(b), (c) and (d) show reducing lengths of the vertical antenna sections and corresponding increases in the lengths of the horizontal components. The total height of the antenna is therefore lowered and in (d), where only 25% of the quarter-wave is vertical, the antenna is only 0.06 wavelength above ground.

The three 'bent' quarter-wave antennas shown in (b), (c)

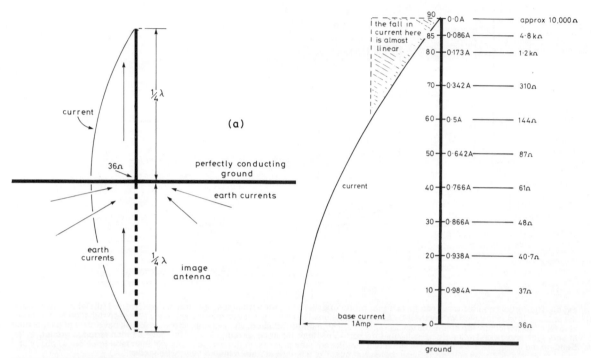

Fig 51. (a) The basic Marconi quarter-wave vertical antenna positioned over 'perfect ground', showing its earth image. Most of the earth-return currents flow through the ground in the vicinity of the antenna. (b) A representation of a quarter-wave vertical antenna over perfect ground which is energised by 36W of power. The RF current at 10° points along its length is shown and also the impedance at these points. There is a rapid fall in current towards the top of the antenna and the impedance therefore rises greatly there. It is interesting to note that the fall in current over the final 30° of this antenna is almost linear

and (d) are often called 'inverted-L' antennas, and they are very popular arrangements when mast height is limited. As the vertical part of an inverted-L is reduced in length, the proportion of the radiated power at low angles and in the vertical plane also diminishes. The horizontal top section will then contribute more of the total radiation, this radiation being horizontally polarised and at high angles to the horizon. This high-angle radiation is a result of the antenna being close to the ground.

An inverted-L similar to that shown at (c), where the vertical and horizontal portions are equal in length, should give useful vertically polarised radiation at low angles for both DX work and also 'local' working within the ground-wave range (normally up to 40 miles). The high-angle radiation from its top horizontal half will be effective for reliable after dark 'first-skip' communication up to a range of about 800km.

In Fig 52(e) the top half of the quarter-wave is dropped down towards the ground. This will reduce both the radiation from the vertical section and also from the sloping part of the antenna, as the RF currents in the two sections will be partly in opposition. In (f) there is both vertically and horizontally polarised radiation, in proportion according to the slope of the antenna. The sloping wire, even halfway along its length, is only 0.05 wavelength above ground.

This antenna will tend to behave as a 'lossy' transmission line, having the ground as a conductor, and any radiation will be at high angles. There will also be very little ground wave, so such an arrangement will be poor both for local working or for DX.

Another feature of inverted-L antennas is that they tend to radiate most strongly in the direction away from the 'elbow'. This directivity can sometimes be useful when particular countries or continents are the radiation 'targets'.

All the antennas shown in Fig 52 are full-sized, quarter-wave Marconi types and they will have a base feed impedance of around 36ohms at resonance. Perhaps the easiest way to arrange a feed to match this impedance is by using two equal lengths of 75ohm coaxial feeder, which are connected in parallel (Fig 53). The author has used this method with considerable success, although for very long cable runs it might prove to be expensive.

The 37ohm feeder must be connected to the equipment via an ATU which will match it to the required 50ohm impedance, and additionally will provide another tuned circuit between the equipment and the antenna. Another way to feed a quarter-wave antenna is to arrange for a tuned circuit at the foot of the antenna which can then be adjusted to give an output impedance suitable for whatever coaxial feeder is to be used. However, such outdoor matching units

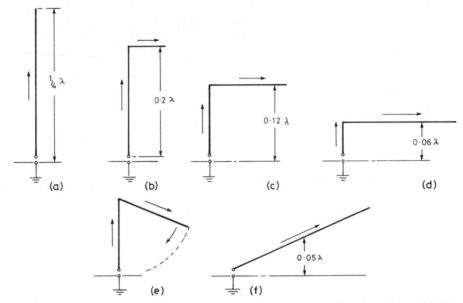

**Fig 52.** The vertical quarter-wave can have a proportion of its length bent horizontally as shown in (b), (c) and (d). When this is done the antenna is usually called an 'inverted-L'. As the proportion of the vertical section falls the vertically polarised radiation at low angles also falls, the horizontal top giving horizontally polarised high-angle radiation. The example shown at (d) will have most of its radiation at very high angles and will only be suitable for short to medium distance working. It will also have a much reduced ground wave. Bending the top of the inverted-L down will mean that the antenna currents in the two sections will then tend to be out of phase and begin to cancel. At (f) the sloping wire will behave almost like a length of unterminated open-wire feeder

need good weatherproofing and they cannot easily be remotely tuned.

## Radiation resistance

In earlier chapters *input impedance,* which is almost a pure resistance seen at the terminals when an antenna is near resonance, and also *reactance,* which is also present when an antenna is either too long or too short to be resonant, have both been discussed. When considering short Marconi antennas, which must be brought up to resonance by some loading technique, either at the antenna top, its middle or at its base, their *radiation resistance* becomes important.

'Radiation resistance' is not a true resistance but is an entirely fictitious one, which, when multiplied by the square of the antenna current, measures the power radiated by an antenna. The energy supplied to an antenna is dissipated either as radiated power or as heat losses in the resistance of the system and in dielectrics in its vicinity.

The total power dissipated is represented by the expression

$I^2R$; in the case of any heat losses $R$ is actual resistance but, when considering the power lost (usefully!) by radiation, $R$ must then become an assumed resistance. This assumed resistance is the 'radiation resistance'. The total power loss in an antenna is

$$I^2 \times (\text{Radiation resistance} + \text{Loss resistances})$$

These loss resistances include earth resistance and the ohmic resistance of the antenna. The radiation resistance of a quarter-wave Marconi antenna is a very important factor in determining its overall efficiency.

## Base loading and top loading

Quarter-wave Marconi antennas can be reduced in physical length and will still radiate if properly loaded to resonance. There are two commonly used types of loading: bottom loading and top loading. There is also centre loading as an intermediate form but this is difficult to arrange, an important exception being in the design of very

**Fig 53.** Two equal lengths of identical 75ohm impedance coaxial cable may be connected in parallel as shown to give a characteristic impedance of 37ohms, which can be used to feed Marconi quarter-wave antennas

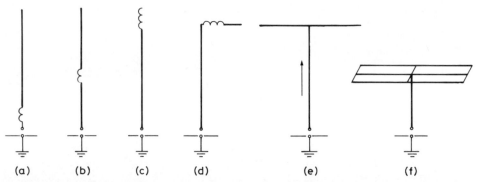

**Fig 54. Short (under a quarter-wavelength) Marconi antennas can be brought to electrical resonance by either bottom loading (a), centre loading (b) or top loading (c) using some inductance in series. By having the loading inductance a little way down from the top of the antenna, as shown at (d), there is a reduced likelihood of corona discharges and the inductance can also be smaller. The 'T' antenna at (e) is an example of top loading, and the fact that the currents in each leg of the top are anti-phase will mean that there will be little radiation from the top. An increased amount of top-loading capacitance as shown in (f) will allow quarter-wave resonance to be obtained with a shorter vertical section**

short antennas for mobile operations. Centre loading is rarely used with fixed-station antennas.

The arrangement shown in Fig 54(a) is a typical example of base loading. The inductance is at the bottom end of the antenna and it makes up the length deficiency of the radiator. This method of loading is the simplest to use with a short Marconi antenna because the inductor can be located inside the shack.

Unfortunately base loading is the least efficient method and it lowers the radiation resistance of the antenna. The maximum radiation from a quarter-wave antenna is at its high-current point, and when base loading is used this will be along the loading coil. When this coil is located indoors the maximum radiation will be there too!

An inductor for base loading will generally be made with a length of wire which is about twice the 'missing' length of the antenna, and this will introduce ohmic losses into the system even when thick wire is used. The author's own Marconi antenna uses no loading coils and its DC resistance, although more than 200ft in length, measures little more than 0.5ohm (16SWG copper wire has a resistance of 4ohms per 1000ft). Note that the RF resistance will be somewhat higher.

Fig 54(b) shows the centre-loading coil arrangement and at (c) an inductor top loads the antenna. This method is seldom used as shown because the inductance must be quite large, and the very high RF voltages generated at its top end may induce corona discharges unless special precautions are taken to stop this.

Fig 54(d) shows a combination of both inductive top loading and some additional capacitance. Extra capacitance at the top of an antenna will allow the use of a much smaller top-loading coil.

At (e) the top-loading arrangement uses just additional capacitance and does not have an inductor. This type of antenna is often described as a 'T', and it is frequently made from a centre-fed Hertzian antenna which has both its feeder wires 'strapped' together at the bottom. The G5RV or tuned-doublet antenna used in this way can become a useful Marconi antenna if tuned against a good ground system. Many intercontinental contacts have been achieved on 1.8MHz when simple 'T' antennas of this type were used.

By employing more top capacitance the vertical section of the antenna may be further reduced in length, as seen in (f). Here a multi-wire, top-loading capacitance is shown. Similar arrangements used with either 'T' or inverted-L antennas have long associations with ship-borne radio installations where there are obvious antenna length restrictions for LF work.

The RF currents in the two horizontal wires at the top of a 'T' antenna are anti-phase, and therefore there will not be much radiation from them in the horizontal plane. Any radiation there is will tend to be at right angles to the line of the wires making up the top. For a more complete suppression of the horizontal component, another pair of wires should be run out at the junction of the 'T' but at right angles to the other two top wires. This can be arranged by drooping slightly the top-loading wires with little loss in the efficiency of the vertical section.

All the antenna variations shown in Figs 52 and 54 are base fed with low-impedance coaxial cable. They are essentially single-band antennas and their feed impedances will be very high on harmonic frequencies; on these high frequencies they cannot be used with low-impedance coaxial cable.

## Efficiency

Antenna efficiency is equal to the power radiated divided by the power supplied, and can be defined as the ratio of the radiation resistance to the total antenna system resistance. Therefore:

$$\text{Efficiency} = \frac{\text{Radiation resistance} \times 100\%}{(\text{Radiation resistance} + \text{Loss resistances})}$$

It can only be 100% when the ground resistance and

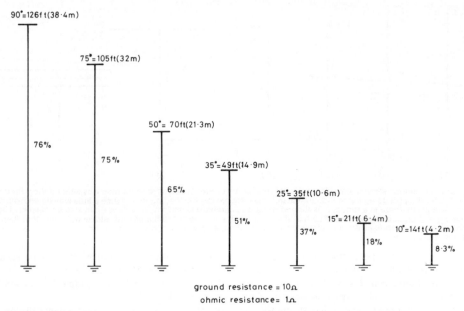

ground resistance = 10Ω
ohmic resistance= 1Ω

**Fig 55. How the efficiency of top-loaded vertical antennas on 1.8MHz falls when they are reduced in length. The efficiencies shown will apply when the ground resistance is 10ohms and the system ohmic resistance is 1ohm**

ohmic antenna resistances are zero. This is obviously impossible! However, efficiencies may begin to approach 100% in the case of an antenna made from thick low-resistance conductor wire or tube situated over an almost-perfect conducting plane (or sea water).

The May 1983 issue of the American publication *Ham Radio* contains an important article by W J Byron, W7DHD, on the subject of short quarter-wave antennas, entitled 'Short vertical antennas for the low bands'. The author is indebted to W7DHD for his computer-generated calculations showing the radiation resistances of base-loaded and top-loaded quarter-wave antennas. These are listed in Table 10, where it will be seen that if a Marconi antenna length is reduced to less than 35° (0.1 wavelength) the radiation resistances when top loaded will be four times greater than those of equivalent-sized but base-loaded antennas.

The power supplied to a quarter-wave Marconi antenna is dissipated in three main ways: in the ground resistance, in the antenna radiation resistance and in the actual and usually small ohmic resistance of the system. To realise high efficiency the greater proportion of the total power must be dissipated in the radiation resistance, so therefore the ground resistance and ohmic resistance must be made as small as possible.

Assuming that there is a ground resistance of 10ohms (which is good and achievable by amateurs) and an ohmic resistance of 1ohm (easily achieved), the different efficiencies of top-loaded verticals ranging down in length from a full quarter-wave (90°) to only 10° will be as shown in Fig 55.

The efficiency only drops off seriously when the an-

tenna's vertical length is below 35° (0.1 wavelength), and fortunately on the 1.8MHz band this length can be achieved when using a 50ft (15m) support mast. Fig 56 shows the same antenna lengths as seen in Fig 55 but using base loading. Now the efficiency of a 35° vertical falls to only 24%, and the very short (about 15ft on 1.8MHz) 10° antenna becomes only 2.4% efficient.

**Table 10. Radiation resistance of short vertical antennas according to W J Byron, W7DHD (90° = quarter-wavelength)**

| Height (°) | Base loaded (Ω) | Top loaded (Ω) |
|---|---|---|
| 90 | 36 | 36 |
| 85 | 30.2 | 35.7 |
| 80 | 25.3 | 34.9 |
| 75 | 21.1 | 33.5 |
| 70 | 17.65 | 31.78 |
| 65 | 14.61 | 29.57 |
| 60 | 12.0 | 27.0 |
| 55 | 9.75 | 24.15 |
| 50 | 7.82 | 21.12 |
| 45 | 6.17 | 18.0 |
| 40 | 4.76 | 14.87 |
| 35 | 3.57 | 11.84 |
| 30 | 2.58 | 9.0 |
| 25 | 1.76 | 6.42 |
| 20 | 1.11 | 4.21 |
| 15 | 0.62 | 2.41 |
| 10 | 0.27 | 1.08 |
| 5 | 0.06 | 0.27 |

The radiation resistance of a Marconi antenna cannot be raised to a value higher than the figure which is applicable

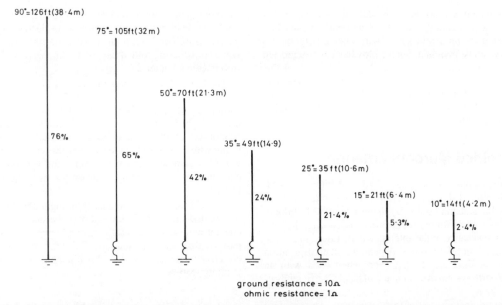

Fig 56. The efficiencies of shortened vertical antennas when they use bottom loading. The top-loaded versions can be more than three times as efficient when the antenna length is small (15° or less, ie 15ft on 1.8MHz)

for the particular antenna length and loading method, so to increase the radiation efficiency the earth resistance must be brought down to as low a value as possible. Earthing systems will be explored in some detail later in this chapter.

The radiation resistances of vertical loaded quarter-wave antennas fabricated from thick tubing or built as a metal tower structure will be a little higher than those obtained when using wire conductors, so in such cases the efficiency will be improved.

## The ³/₈-wavelength Marconi antenna

Quarter-wave inverted-L antennas unfortunately have their greatest radiation down near their feedpoints, with the current maximum being positioned very close to or even

inside the house. The dielectric and screening effects of a building can seriously reduce such an antenna's performance.

To help to overcome this, the author lengthened one of his inverted-L antennas from a 1.8MHz quarter-wavelength to about 190ft (57.9m), which is a ³/₈-wavelength (see Fig 57). The antenna's RF current maximum was then moved from ground level to a point some 126ft (38.4m) from the far end of the antenna, and was additionally raised to about 45ft (13.7m) above the ground.

Although this is not high in terms of wavelength when on 1.8MHz, the new elevated high-current point really improved the overall performance of the antenna as compared with the quarter-wave 'L', and many contacts with VK6 and other DX stations were made.

Amateur operators often have only modestly sized

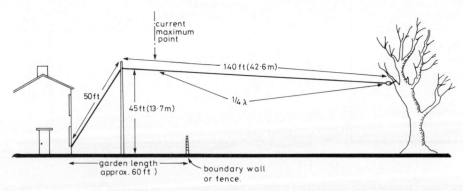

Fig 57. An end-fed inverted-L which is ³/₈-wavelength long on 1.8MHz. This is a convenient length because it puts the high-current point high up and away from the house, and it will also be easier to match than an end-fed half or quarter-wavelength of wire

gardens which will not accommodate an antenna of this length, but sometimes, and with some gentle persuasion, permission to run a thin single wire over a neighbour's property can be obtained. Such a ploy has been suggested in Fig 57! The impedance at the shack (feed end) of this antenna will be higher than the 36ohms of a quarter-wave wire so an ATU must be used. An added bonus is that the antenna will work well on all bands and on the higher frequencies will be an effective long wire.

## The folded Marconi antenna

This antenna type has been described by Bill Orr, W6SAI, who has offered slightly differing versions of it in at least three of his antenna books. By using a length of 300ohm ribbon as a 'half-folded dipole' or 'folded unipole', the radiation resistance of the antenna will be raised to a value in the region of 80-150ohms, depending on configuration and height. This means that even when used with an average earth system the antenna will have an efficiency of around 40%.

In the original version of this 'twin-lead Marconi' (see *S9 Signals* by W6SAI) ordinary 300ohm ribbon was used for the full quarter-wavelength of the antenna, but a later version, which was described in his book *Simple Low-Cost Wire Antennas* (published by Radio Publications Inc in 1972), has a shortened 300ohm ribbon section to which is added a length of single wire.

To calculate the length of the ribbon needed, an electrical quarter-wavelength at the desired frequency must be multiplied by the velocity factor of the ribbon used (slotted ribbon has a velocity factor of 0.87). This length will be

short of a quarter-wave, and it must have an additional wire connected at its end to make it up to be an electrical quarter-wavelength. This technique is very similar to that used when constructing folded half-wave dipoles from 300ohm ribbon (see Chapter 2).

The folded Marconi illustrated in Fig 58 is designed for 3.7MHz operation and it only needs 35ft (10.6m) high supports. The antenna's efficiency will be proportionally higher as the length of its vertical section is increased as a proportion of the total quarter-wavelength. Six buried radial wires, each being at least a quarter-wavelength long, will provide a suitable earth system, and versions of this antenna may be scaled up or down for working on other bands.

The step-up of feed impedance brought about by using this folded dipole technique allows the use of a 50ohm coaxial feeder. The greater distance between the vertical part of this antenna and any buildings etc, the more effective the antenna will be for low-angle long-distance communication.

## A three-wire sloping Marconi *et al*

The folded Marconi antenna just described has an impedance (and radiation resistance) step-up of four times, but if three quarter-wave wires are used this will increase to a factor of nine times. This raises the feed impedance to about 300ohms, and the antenna can then be fed with a length of 300ohm ribbon (or via a 4:1 balun to 75ohm coaxial cable).

The use of such a multi-wire radiating element will also lower the Q and increase the bandwidth of the antenna.

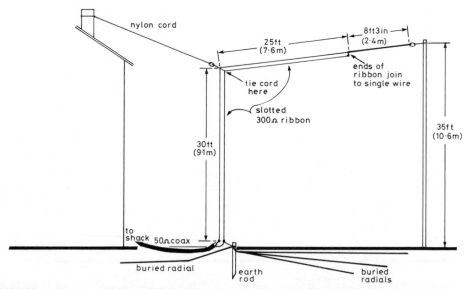

Fig 58. The folded Marconi antenna which uses 300ohm ribbon for most of its length. The use of the 'folded dipole' principle raises the feed impedance of this antenna from around 15ohms to four times this figure. A reasonable match can be obtained with 50ohm coaxial feeder

This means that its feedline SWR will remain low for a considerable range on either side of the its resonant frequency.

The three wires must be spaced about 1ft (30cm) apart and, as they use an air dielectric, their lengths can be calculated from the standard formula used to find the length of a quarter-wave: $234/f(\text{MHz})$ feet (which must be multiplied by 0.3048 to convert to metres).

Sloping this antenna is the most convenient arrangement – so long as the slope angle does not exceed 30° it will perform almost like a true vertical and provide low-angle radiation. The version shown in Fig 59 is designed for the 3.5MHz amateur band but a half-sized version would be fine for the 7MHz band – only a 30ft high single support would be needed. The spreaders S can be fashioned from lightweight ¹⁄₄in. (18mm) diameter plastic pipe which is sold in DIY shops as 'plastic waste pipe'.

Quarter-wave antennas made as folded 'half dipoles' will present similar feed impedances at both their fundamental frequency and at their third harmonic frequency. An antenna of this type, when resonant on 7MHz, will also work on its third harmonic of 21MHz, but then of course it will have the radiation characteristics of a vertical ³⁄₄-wave antenna.

## Grounded Marconi antennas

When one end of a single-wire antenna is grounded, that point will always remain at earth potential, and that end of the wire will also carry a high RF current. This technique is not new but has been little used in amateur antenna design. R Cornet, PA0RCH, published a design for a top-loaded, grounded 'half-quad' in the August 1976 issue of the Dutch magazine *Electron*, and the work of Belrose, VE2CV, in developing his grounded half-delta antenna is described in Chapter 4. The author has a 40-year-old QSL card from AC4RF (Tibet) which describes his antenna as ''...a grounded long wire...''.

The author's experience using grounded antennas dates back more than 20 years (see *RSGB Bulletin* for February 1964) when he sloped a quarter-wavelength of wire down to the ground from his shack window, which was near the top of a tall Victorian villa. This wire was resonant on 3.7MHz and its lower end was well grounded to an ancient buried metal water tank and several buried wires. The top of the sloping wire connected to a very short (6ft) length of open-wire feeder which terminated at the station ATU.

This simple and rather crude vertical quarter-wave antenna worked well, and for the first time the author could contact VK and ZL stations when using a home-brew 50W PEP transmitter on 3.5MHz. The idea was revived a few years ago when experimenting with antennas for long-distance working on 1.8MHz which would also radiate on the other HF bands. Monoband antennas are fine for the specialist with lots of available garden space, but for ordinary mortals who are restricted to the confines of the average-sized garden they represent a luxury!

To obtain the low angles of radiation needed for successful DX work on the LF bands, the choice lies between very high Hertzian antennas or instead Marconi antennas, which will provide vertically polarised low-angle radiation.

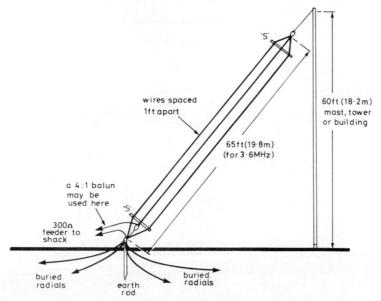

Fig 59. The three-wire sloping quarter-wave Marconi has an even greater impedance (and radiation resistance) step-up. This design is for a 3.6MHz version and it has a feed impedance of about 300ohms. A 4:1 balun can be used at the feedpoint and then 75ohm coaxial cable may be run back to the operating position. The increased radiation resistance means that the antenna will work well even with a high-resistance ground system

Inverted-L Marconi antennas, when about a quarter-wavelength long, unfortunately have the maximum radiation down at their feedpoint which is often close to or inside the house. An inverted-L can instead be rearranged so that its 'elbow' is away from the house and its vertical section comes down to the ground where it is effectively earthed. The horizontal top can be made to run back to the station ATU and its end becomes the feedpoint of the antenna.

This 'backwards' inverted-L antenna will have certain advantages over the more conventional arrangement. There will be vertically polarised radiation from the vertical section, and the top of the antenna becomes both a feeder and a loading section.

The horizontal part will radiate, but this can be a bonus for it can provide some high-angle radiation for short-distance working. An antenna arranged like this can also be put to good use on any of the higher-frequency bands; it will not become neglected or redundant during the summer months when the LF bands are often (and mistakenly) ignored. Fig 60(a) shows a full-sized quarter-wave vertical antenna having an additional horizontal top feedline which is less than a quarter-wavelength long. Such an arrange-

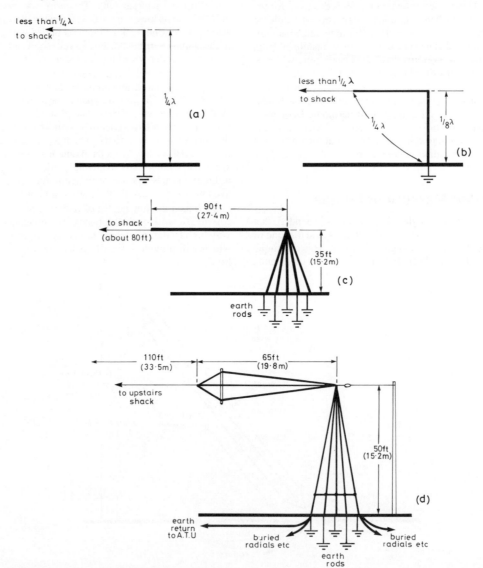

Fig 60. (a) A grounded, quarter-wave Marconi antenna which is top fed by a single wire. (b) A similar antenna but with only an eighth-wavelength in the vertical plane. (c) Here the vertical part of the antenna is little more than $1/16$-wavelength but is made with several wires. (d) The 'steeple' antenna which is essentially the antenna in (c) but having some additional top loading. In many ways this antenna resembles the grounded half-delta

ment would be best adapted for the 7MHz band, for then it would only need to be up to a height of about 35ft (10.6m). At (b) the vertical section is reduced to an eighth-wavelength (which on 1.8MHz is still about 63ft or 19.2m), but the horizontal top is a further eighth-wavelength plus a feeder section which should again be shorter than a quarter-wavelength. By having the feed section under a quarter-wavelength long, the RF current at the feedpoint will not be high and the radiation there will be reduced.

The scheme shown at (a) has the disadvantage that the vertical section (when used on 7MHz) will be only about 20ft (6m) from the house and will be screened in that direction.

The author only had a 35ft mast available when his tests with grounded wires began; the arrangement shown in Fig 60(c) was tried. Five wires were brought down to separate earth rods which were arranged in a 6ft square pattern in an attempt to reduce the earth resistance and also to 'broadband' the antenna by bringing down its inherent $Q$. The top wire was 170ft (51.8m) long, the first 90ft (27.4m) of which, when measured from the top of the vertical section and added to this, made up a quarter-wavelength on the 1.8MHz band.

The junction of this 90ft length of wire with the remaining 80ft (24.4m) of feed wire (actually it was all one wire!) is the high-voltage end of the quarter-wave. From this point back to the feedpoint in the shack the RF voltage falls and the RF current increases to a median value. This means that right at the feedpoint there will be a medium impedance with some capacitive reactance, which should not present any matching difficulties when using an ATU.

This antenna worked well on 1.8MHz and a lot of DX was heard and worked on that band. It could also be tuned up easily with an ATU on all the other HF bands, where it had a creditable performance.

## The 'steeple' antenna

Shortly after the acquisition of a sectional 50ft aluminium mast the antenna shown in Fig 60(d) was designed and erected. Its vertical section had eight slightly sloping wires which all went to individual earth rods, so giving the general appearance of a church steeple – hence its name!

By using a three-wire 65ft (19.8m) horizontal section, there was a considerable capacitive top loading of the vertical 50ft (15.2m) grounded-wire section. The three wires behaved as a capacitance 'hat', and each 11 sq ft of their area contributed about 40pF of capacitance. The area of the loading section was around 200 sq ft so its total capacitance was about 700pF.

When using shortened versions of such a top-loading section, it is as well to remember that if all its linear dimensions are reduced to 71% of the shown size there will then be only half of the original area, and the top capacitance will fall to 350pF.

Plastic $^3/_4$in. (18mm) diameter 'waste pipe' was used for the spacers and each one was 6ft (1.8m) long. According to Byron, W7DHD, this antenna should have a radiation resistance of 11.8ohms. To achieve a reasonable radiation efficiency with this radiation resistance, a very good low-resistance earth system is essential, so a considerable amount of time and energy was directed towards that end.

A spring gale revealed an inherent weakness in the system, for several of the plastic-covered stranded wires used to make up the antenna snapped. Three of these were at their connections to the earth rods. Before this calamity the antenna had given a good account of itself on 1.8MHz and also the higher-frequency bands, and it was noticeably more effective than its prototype in Fig 60(c).

John Belrose (VE2CV of grounded half-delta loop fame) read the author's description of the 'steeple' in *Radio Communication* (August 1986), and wrote informing him that this antenna was really a half-quad loop. The total length from its feedpoint to the ground connections is about 225ft (68.6m), which is a little short of a half-wavelength on 1.8MHz. The length of the earth-return wire between the upstairs shack and the ground makes up this length deficiency.

John is probably correct in his assumption, but experiments carried out by the author and others, which involved using a greatly reduced length of top section (down to under 100ft or 30m), seem to demonstrate that when the total length from end to end is little more than a quarter-wavelength the 'steeple' will still work quite efficiently. As some folk say, ''You pays your money and takes your choice...''!

## The 'double-bass' antenna

The author fortunately has a large garden in a rural area, and also has friendly and tolerant neighbours, so he was able to replace his 'steeple' with the 'double-bass'! This name was decided by the number of strings on that instrument, its size and the fact that in high winds the antenna wires produce strange musical sounds rather like an aeolian harp. Such an antenna is obviously not suitable for use in a small urban garden, for there its 600ft (183m) of airborne wire would most certainly not 'melt into the background'!

### Construction

The success of the 'steeple' seemed to come from the addition of its top-loading section, so this led to the idea of having an even greater top-loading capacitance. The arrangement shown in Fig 61 was decided upon and built. Four 100ft (30.5m) lengths of hard-drawn copper wire were tied to a convenient anchor point near the bottom of the author's garden, and at their halfway point a 6ft plastic spacer was fitted.

Similar spacers were also fixed at points 10ft (3m) from

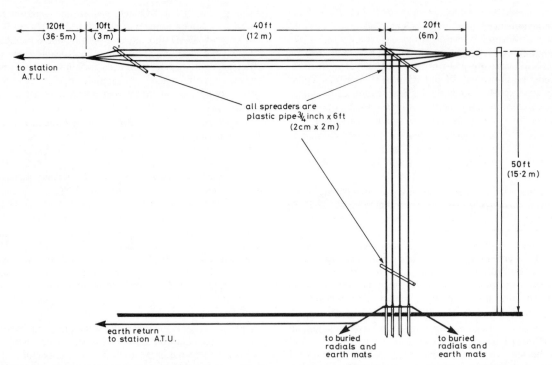

Fig 61. The 'double-bass' antenna is an improved version of the 'steeple' and it has considerably more top-loading capacitance. This antenna (like the 'steeple') also works well on the higher-frequency bands

both ends of the wires. The four wires were then brought together at one end and soldered to the 120ft (36.5m) single-wire 'feeder' which led to the upstairs shack. Four additional wires each 20ft (6m) long were soldered to the halfway points along the 100ft wires. These made up an additional top capacitance and, as part of an unbalanced 'T', would also help to reduce some of the horizontally polarised radiation from the top of the antenna.

Instead of arranging the earth rods as a square, as was done with the 'steeple', they were instead set out in a line measuring 6ft end to end. The tops of these earth rods were connected together with a thick conductor wire made from four lengths of 16SWG copper twisted together, to which the bottom ends of the four vertical wires were soldered. Thin but strong nylon cords were tied to the outer ends of the plastic spacer (which was about 10ft from the ground), and these were then taken down to small ground stakes. These cords restrained the sway of the vertical wires in windy conditions and took away a lot of the mechanical stress from the soldered connections.

The total top capacitance (calculated from the area of the loading section) is around 1500pF, which is enough to load the 50ft wires to near an electrical quarter-wavelength on 1.8MHz. The use of four wires for both the antenna vertical section and the top-loading section increases the system's bandwidth, lowers its ohmic resistance and, more importantly, raises the radiation resistance. It has some of the characteristics of a multi-wire folded unipole.

Fig 61 also shows an earth-return wire which runs from the base of the antenna back to the station ATU. This wire can be buried or, as has been done by the author, part buried, and then run along and inside a hedge before running up the house wall to the upstairs shack. The success of this antenna and others of similar design will depend upon the effectiveness of the earthing arrangements, and suitable ground systems will be explained in some detail later in this chapter.

## Performance

Although primarily designed for long-distance working on the 1.8MHz band, the 'double-bass' antenna has additionally proved to be very effective on the higher-frequency amateur bands, especially 21 and 14MHz. On these bands it seems to out-perform many of the tri-band 'trapped' antennas, using inductors, and some rotary beams that are in common use.

It does not display any noticeable directional properties on any band and, as it is an earthed system, it is not prone to static-charge build-up. The fact that both the 'steeple' and the 'double-bass' are always grounded will inspire some small feeling of confidence during electrical storms! Closed-loop systems are always less noisy than open-wire arrangements, and on the LF bands this is an important factor in their favour.

Comparison checks of received signal strengths on all

bands between 3.5 and 28MHz, when using either the 'double-bass' antenna or a commercially made trapped multiband vertical, show that the big antenna has at least a two, and often a four, S-point advantage. This advantage represents from 12 to 24dB gain, which is very considerable indeed, although the trapped vertical antenna is ground mounted and not in the clear.

The 'double-bass' unfortunately picks up a lot of electrical noise from properties in the surrounding area, a particular nuisance being an electric cattle fence. The author also has an inn only 100m away which has many electrically operated mixers etc – these can be the source of much QRM on the 1.8MHz band. To overcome this problem a couple of screened loop antennas (see Chapter 6) have been put into the roof space above the shack, and these are often used for reception on the 1.8MHz band.

## Earth systems for Marconi antennas

Unless ship-borne or living in the middle of a salt marsh, a major problem facing the user of Marconi antennas is the provision of a really effective earth system. Contrary to what is often generally accepted by many radio amateurs, the earth is not really a conductor at all, but a lossy dielectric! Earth conductivity is dependent upon the frequency being used, and what is considered a 'good earth' by electricity supply engineers may be useless at RF.

An American naval study of earth resistance produced some illuminating facts, and a brief look at some of its findings is revealing. If sea water is assumed to have a relative conductivity of 4500, the relative conductivity of the *best* moist rich soil is only 15. For average soil this falls to 7 and on sandy or chalky soils the conductivity drops still further to a value of only 2! In built-up urban areas the value is also 2 but in city industrial areas it is down to only 1. Fresh water such as is found in lakes, ponds, rivers or streams has a conductivity value of 6, which is lower than that found in average soils! Having a pond or a stream or a high water table in a garden is (it seems) of no advantage.

The earth must therefore be considered as a lossy dielectric and not as a conductor. Such a dielectric beneath a Marconi antenna will introduce losses, and it is this fact which is responsible for the poor performance of such antennas when only a few counterpoise wires or buried radials are used. Much of the earth-return current just cannot find its way back to the base of the antenna and the earth resistance will be high.

Table 11 reveals the importance of having a low-resistance earth system when Marconi antennas are used. A ground resistance of 100ohms, which is not untypical of many simple earth systems used by radio amateurs, will reduce the efficiency of a full-sized quarter-wave vertical antenna down to 26%. This climbs up to 76% when the earth resistance is brought down to 10ohms. This is attainable by amateurs and, if even greater effort is made and the earth resistance is reduced to 5ohms, the efficiency will rise to 85%.

**Table 11. Ground resistance and the efficiency of vertical antennas**

| Antenna height (°) | Efficiency base loaded (%) | Efficiency top loaded (%) |
|---|---|---|
| Ground resistance = 100Ω, ohmic resistance = 1Ω | | |
| 90 | 26 | 26 |
| 50 | 7.1 | 17 |
| 10 | 0.26 | 0.98 |
| Ground resistance = 10Ω, ohmic resistance = 1Ω | | |
| 90 | 76 | 76 |
| 50 | 42 | 65 |
| 10 | 2.4 | 8.3 |
| Ground resistance = 5Ω, ohmic resistance = 1Ω | | |
| 90 | 85 | 85 |
| 50 | 56 | 77 |
| 10 | 4.3 | 15 |

When shorter and loaded verticals are used their efficiencies over a poor earth will be very low, and in the case of a 50° base-loaded antenna will be only 7%. This can be raised by eight times when the earth resistance is reduced from 100ohms to 5ohms.

## Earth rods

The use of earth rods by amateurs can be traced back to the 'twenties and 'thirties when most of the radio magazines extolled their virtues. They often illustrated 'ideal' systems using an inverted-L antenna and a copper earth rod driven a few feet into the ground just outside the nearest window. It was often suggested that in dry weather the soil around the earth rod should be watered, and some authors even advocated the use of saline solutions!

If earth rods were effective when used with Marconi antenna systems they would have been set out by many commercial and broadcast stations. Several hundreds of these rods all bonded together would have been much cheaper than the extensive copper ground mats employed at such installations.

Bill Byron, W7DHD, who is an expert antenna design engineer, informed the author by letter that "...earth rods are virtually useless as ground for the radio frequencies we use today. They are, however, very useful for the absorption of lightning discharges – up to a point...it is also pretty well known that for whatever use they are put to there is little benefit in spacing them closer than their own length. In other words, their effectiveness is not proportional to their exposed surface area. It is proportional to their length, and here again I am talking about low frequency lightning discharges...''.

All amateur radio installations should have at least one earth rod, but only as some small precaution against minor lightning discharges. Earth rods must be discounted as being of any value for earthing when using Marconi antenna systems. The rods specified earlier in this chapter are merely anchor points for vertical wires or connecting points for additional parts of an earth system, and they contribute very little to the proper working of the antennas described.

## Counterpoises

An interesting article appeared in *QST* magazine in February 1983, entitled 'Efficient ground systems for vertical antennas', which examined the case for and against counterpoise systems and buried radial wires. This was the culmination of a study made by a small group of 'old timers', K8CFU, W3ESU and K4HU, and their experiments revealed that a proper counterpoise system was far more effective than buried radial wires.

Their earth resistance measurements showed that there were very great variations all over their 300 by 200ft (91 by 61m) test site. These changes in resistance were quite dramatic and often took place, for no obvious reason, at points only separated by a few feet. They discovered that such variations in the earth resistance caused variations in the levels of the return currents flowing in individual buried radial wires. However, when elevated counterpoise wires were used (insulated from the ground at their ends and along their lengths), the earth resistance variations then had a much reduced effect upon the return currents.

A counterpoise must be elevated to between 6 and 10ft (1.8 and 3m) above the ground, and this is a major obstacle when planning an amateur installation. If some unused ground is available an elevated counterpoise system of just 32 wires arranged within a 100ft (30m) square will be very effective when used with Marconi antennas on 1.8MHz.

Such an arrangement was tested and named the 'minipoise' by the three amateurs who made the study (see Fig 62). Unfortunately, few amateurs in urban or semi-urban environments have enough space to set up a minipoise so they are obliged to turn to other methods in the reduction of their ground resistance.

## Buried radials

One great advantage of using buried radial wires rather than a counterpoise system is that they cannot be seen! It is generally accepted that the greater the number of buried wires used, the more effective the earth system will become, and some amateurs spend much time and money in 'planting copper'. Many who use the 1.8MHz band for DX working show details of their antenna systems on their QSL cards. Two such cards received by the author are illuminating: "...Vertical with 12,000ft (2½ miles) of radials" – K5UR; "...Two-element vertical tower array with six miles of radials" – K6SE.

A leading European 1.8MHz station, SM6EHY, is reputed to have 26km (more than 16 miles) of radials! Many British stations also have extensive radial systems, and one was recently heard saying that he used 200 radial wires, all

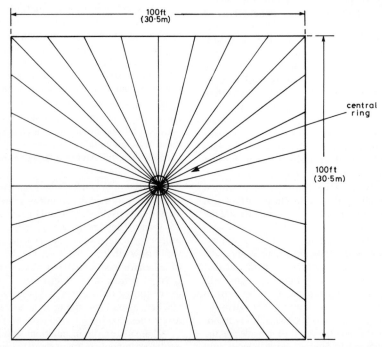

**Fig 62. The 'minipoise' counterpoise arrangement, devised by a small group of 'old timers' in the USA, which proved to be very efficient when used with a short, top-loaded 1.8MHz antenna**

of which were at least a quarter-wavelength long. It is not unknown for amateur stations to use sets of buried radial wires set only 1° apart, but for the majority who have only small or medium-sized gardens such earth systems must remain a dream.

A minimum of six radial wires is a good starting point, and additions to this modest arrangement can be made later when time and resources allow. The wires do not need to be cut to a resonant length, for a good earth system is not a ground plane, and any length of wire 20ft (6m) or longer is useful.

In Fig 63 some different radial spacings have been drawn in the quadrants of a circle. In the southwest quadrant are what must be a minimum number of wires which are spaced to be 45° apart.

Of the four arrangements shown, that in the northeast sector is the best. Although this only contains six longer wires spaced at 20° it also has many shorter wires at 5° spacing. These provide a concentration close to the base of the antenna where the return currents are strongest. It is not always possible to lay out the radials symmetrically as shown in Fig 63(a), so instead a much different scheme might be used which adapts to the dimensions and layout of the available ground.

An imaginary but in many ways typical garden plan is shown in Fig 63(b), which echoes some of the layout problems experienced by the author. The disposition of the many radial wires needed may at first seem alarming, but once they are buried all signs of their existence soon vanishes. The aim is to avoid ground which may be cultivated in the future, and yet cover as great an area as possible.

The presence of concrete pathways will prevent wire runs in some directions unless tunnels are made beneath such paths. An aluminium-framed greenhouse in a garden can be bonded to the earth system, as can also any metal fencing wires.

Special ploughs are used when the extensive earth systems of broadcast stations are being put down, but the amateur can utilise a garden spade to make linear slots, which are 2 or 3in. deep, and into which the radial wires may be pushed. A firm tread over these slots afterwards binds together the soil and all traces of disturbance soon disappear.

## Earth mats

The author's garden now has a total of about half a mile of buried wires somewhere below its surface, and it becomes difficult to remember just where each length of wire lies. Several lengths of earth mat made with galvanised-iron 'chicken wire' have been put down to improve the earth-return conductivity in the vicinity of the base of the grounded Marconi antenna. These have made a very noticeable improvement to the overall performance of the antenna, much more than any further additions to the buried radial system seem to make.

Such woven wire strip can be bought in 10m long rolls and is 1m wide. The cheaper 2in. (5cm) mesh is quite adequate for the LF bands, for such a mesh size is very small in terms of wavelength and the 'mat' will perform almost as well as a flat metal sheet.

## Laying the mats

Earth mats of this type can be laid without any digging or disturbance of the ground surface. They must be put down during the growing season of the grass. The first task is to

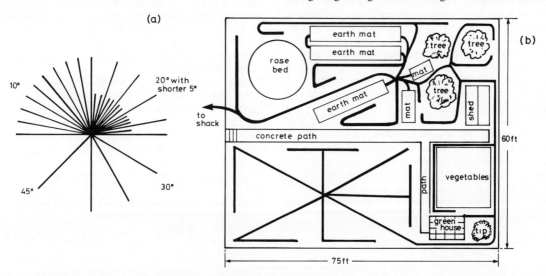

Fig 63. (a) Different spacing arrangements for a buried ground-radial system. (b) How a useful low-resistance earth system can be arranged within the confines of a typical small urban garden

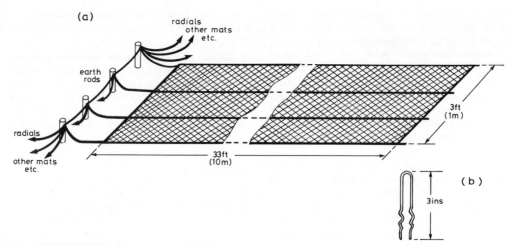

**Fig 64.** (a) The way a length of 'chicken wire' can be laid as an earth mat on a grassy surface. (b) A wire clamp to hold the chicken wire down to the ground. Similar clamps should be used at 1½ to 2ft (45 to 60cm) intervals along the length of the wire and at selected points within its area

make a very short grass cut with a mower along the line of the projected mat. It is best to run over this area of grass several times with the cutter to get it really short.

Fig 64(a) shows how the earth mat is laid down and connected to the central earth point with at least four wires. These wires are soldered to the four thicker strengthening wires which run the length of the chicken wire.

At (b) is shown a 'U' shaped clamp which can be made from 16SWG iron wire. It has kinks along the length of the prongs and these will help retention in the ground. A number of similar clamps can be used at 0.5m (18in.) intervals to hold down the earth mat along its sides and at a few places within its area.

Particular attention must be given to the two ends of the wire mesh roll, and there some longer clamps can be used. After a few days this type of earth mat will begin to vanish from view, and after a few weeks the grass growing over it can be mown. Such mats disappear completely after an interval of about two months and this is brought about by worm action. These creatures bring soil to the surface each night and any object left lying on a grassy surface will eventually be buried.

The author's earth system now incorporates 800 sq ft of wire mesh and this area is added to each summer. If possible it is best to lay the initial lengths radially from the base of the antenna so that they are equidistant. The effectiveness of a vertical Marconi antenna seems to be greatest in those directions where the earthing system is best. An initial three mats spaced at 120° can have additional intermediate sections added at a later date.

## Try all the systems

A combination of buried radials, earth mats and elevated counterpoise wires is perhaps the best approach when arranging a good low-resistance earth system for a small or average-sized garden. Counterpoise wires using plastic-covered stranded wire can be run along fences and right inside growing hedges, and the perimeter of even a small property is a considerable length.

Any such wires must be insulated from the ground and can be from 3 to 6ft above it. They should all connect back to a common earthing point at the base of the antenna. If a property has a separate front garden it is useful to have a second common earth point there with its own system of radial wires etc. This second earth point can connect directly to the chassis or earth connection on the station ATU, with a very thick conductor wire (made with several copper wires in parallel) between the two. Both earth points can be joined together with the long earth-return wire as shown in Fig 61.

Cold-water main pipes can be useful additions to an earth system, and a connection to these will normally connect to the earth point nearest the operating position. Gas pipes (mostly made of plastic these days), central heating pipes and mains earths are best left alone for reasons of safety and the avoidance of electrical noise pick-up.

Improving any earth system is an ongoing exercise, but it will always be reassuring to reflect that each additional buried wire, counterpoise or earth mat will contribute to the radiation efficiency of the antenna.

# Chapter 6

# A gallimaufry of antennas

## The Windom

Loren G Windom (8GZ, 8ZG and later W8GZ, W8ZG) was a prominent American DXer during the 'twenties. For a number of years he held the QRP distance record for his contact with the Australian station A5BG in 1925, when only using an input power of 0.567W. The distance achieved was 10,100 miles which represents 17,820 miles per watt of input power.

However, Windom is not now remembered for his feats of DX, but for an article he wrote in the September 1929 issue of QST which dealt in some depth with a single-wire feed antenna (see page 1). It must be remembered that low-impedance coaxial (or even twin-wire solid dielectric) feeder was not available at that time, and an increasing number of resonant half-wave or harmonic Hertzian antennas were being used with feed at either their ends or at their centres with open-wire, tuned transmission lines.

Most modern antenna books disregard the Windom antenna, for just after the second world war it acquired a poor reputation and was blamed for TVI. Now that the TV transmissions in the UK are on UHF channels, a correctly set up Windom antenna ought to be no more troublesome where EMC is concerned than many other antenna types.

A single wire of either 14 or 16SWG has an impedance of about 600ohms when used 'against ground' (the ground behaving as the second conductor of a balanced line). This allows such a single wire to be connected to a point having this characteristic impedance along a resonant half-wave antenna wire, in order to give a good match.

The impedance along a half-wave (at resonance) from its centre and towards its ends will rise from a low value to one as high as 100,000ohms, and it happens that the 600ohm position is about one-third of the distance from each end (actually 0.36 of the antenna length).

It is more usual to determine the Windom antenna tap point by measuring the distance D (see Fig 65(a)) from the antenna centre point. This distance in feet can be found by the simple formula $D = 66/f(\text{MHz})$ feet. The result must be multiplied by 0.3048 to determine D in metres. A Windom antenna's length can be calculated from the usual $L = 468/f(\text{MHz})$ feet which is correct for a wire half-wave top.

## Some practical points

The basic Windom is a half-wave radiator, so to be effective it must be at least one half-wavelength above the ground. Its single feed wire must be arranged to drop vertically below the antenna for at least a quarter-wavelength, and then it may bend gradually towards the amateur station window. It can theoretically be of any length, for when correctly matched to a resonant half-wave top it will not have any standing waves along its length and therefore it should not radiate.

In practice this is almost impossible to achieve, and the many Windom antennas that the author has used have all radiated from their feedlines to some extent. When using moderate power levels there will be quite a high RF voltage along the feeder, so it must be positioned where any accidental contact with people or animals is impossible. A good earth is essential for the correct operation of a Windom for the earth is one 'leg' of the feeder, so at least one resonant counterpoise wire running beneath the antenna is recommended.

## Setting up

The classic way to set up a Windom antenna was outlined in the earlier editions of *The Amateur Radio Handbook* (RSGB). First, the half-wave top must be trimmed to resonance – to do this a pair of RF ammeters should be inserted into the antenna on either side of the tap point (with the feed wire connected). When the antenna is correctly tuned to resonance by applying some power and adjusting the top length until both meters show the same current, the correct feeder tap point can then be found.

Few amateurs today have suitable RF ammeters so it is suggested that a pair of 60mA 'fuse' lamps may be used in their place. These do not have to be in the actual antenna circuit but can be tapped over a few inches of the antenna wire on each side of the tap. This tuning method implies that bright and sunny days must be avoided! A pair of binoculars can be used to study the bulb illuminations and the exercise is best carried out at dusk.

The correct tap point is also determined by the use of RF current indicators. Again, 'fuse' bulbs shunted across the feed wire, one close to the actual tap and the other about a

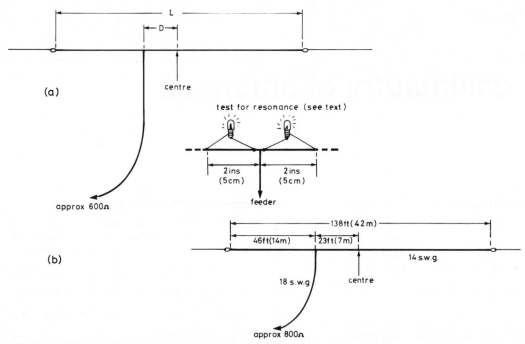

Fig 65. (a) The Windom antenna, which is essentially a horizontal half-wave fed with a single wire which connects at a suitable impedance-matching point. The two 'fuse' bulbs are used to determine resonance of the top before the feeder is permanently connected. (b) The VS1AA multiband version of the Windom. The single-wire feeder is thinner than the wire used for the antenna top and this helps the matching. This antenna uses the one-sixth principle: the distance between the antenna centre and the tap position is one-sixth of the top length

quarter-wavelength away down the feeder, can be used. These can again be shunted over a few inches of feeder wire (both having the same shunt length), and the correct feeder tap point is found when the currents in both the bulbs are equal.

The author's easy way to check the correct tap point is by running a small neon lamp along the feed wire when RF power is applied. The neon should strike and show the same illumination over a quarter-wave of feed-wire length. A neon was taped to a 12ft (3.7m) bamboo pole and this simple 'gimmick', when used with a step ladder, proved to be a very effective voltage indicator.

## The VS1AA multiband antenna

The late Jim MacIntosh (later GM3IAA) was the inventor of the VS1AA antenna. Jim's antenna experiences went back to the first world war, when as a young Army signaller stationed in Egypt he arranged for long wires to run down from the top of the Great Pyramid! He worked for many years in Malaya and it was when he was living in Kuala Lumpur that he devised his variant of the Windom antenna (see Fig 65(b)).

VS1AA's article on this was published in the RSGB *T & R Bulletin* in November 1936, where he demonstrated that, if the feeder tap point was made at a distance from its

centre of one-sixth of the antenna's top length, this would also be correct when the antenna was operated on its even harmonics. Such an antenna, when 138ft long (Fig 65(b)), would work well on the 3.5, 7, 14 and 28MHz bands.

To improve the match, which will always be at a higher impedance than the tap point of a Windom, ie at about 800ohms, MacIntosh suggested the use of a thinner conductor wire for the feeder. If the top wire is made with 14SWG the feeder should be a length of 18SWG wire. A half-wavelength of wire on 3.5MHz is almost 134ft long but the stipulated length for a VS1AA top of 138ft is correct for the even-harmonic operation of the antenna. The antenna will only be 3% away from its true resonant length as a half-wave on 3.5MHz, and this difference will not affect its operation on that band.

The second antenna ever used for transmitting by the author was a VS1AA and, when up in a very indifferent location in 1947 and using under 20W of power output on the 7, 14 and 28MHz bands, much DX was worked. The VS1AA also needs a good earth or some counterpoise wires if it is to work efficiently.

## A coaxial-fed full-wave antenna

Here is another asymmetrically fed wire antenna (Fig 66), but a coaxial feeder is used instead of the single-wire feed system employed for the Windom and the VS1AA. This

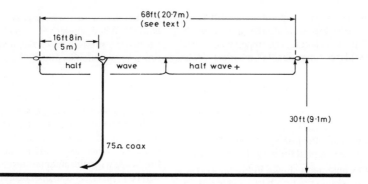

**Fig 66. The coaxial-fed full-wave antenna is a monoband system which has a full-wave top-fed a quarter-wavelength from one end with 75ohm coaxial cable**

antenna, which is a twin half-wave wire, is fed a quarter of a wavelength from one end with 75ohm coaxial cable.

The two half-waves which make up the top length work out of phase so the antenna radiation pattern is similar to the familiar four-lobed display of a full-wave antenna. This antenna will work well on 14MHz and give low angles of radiation when it is only 30ft above the ground.

## Tuning and pruning

To set it up correctly it is best to start with a top 68ft long. This is broken 16ft 8in. (4.87m) from one end (nearest the shack) and here the 75ohm coaxial cable is connected without a balun. The feeder must drop vertically for at least a quarter-wavelength (16ft) to prevent imbalance and feedline radiation.

Using low power, and with an SWR meter inserted in the feedline, readings can be made across the band. If the top is too long, the lowest SWR will be at the LF end of the band so the far end of the wire top (the ³/₄-wave section) can be pruned a little. At this stage it is best to just bend the unwanted wire back on itself without actually cutting it, and the SWR measurements can be resumed.

Several attempts at this 'cut and try' technique will eventually result in a antenna which has a low SWR at mid-band (typically 1.1:1) and which will rise to no more than 1.2:1 at the band edges. John Lunn, G3BRD, tried this antenna and discovered that each inch of wire removed

from the antenna top raised its resonant frequency by about 100kHz. When the correct length has been found the excess wire can then be cut and removed.

The antenna can be extended to include additional half-waves and it will then have the gain and radiation pattern of an equivalent long wire. A meter adjusted for 75ohms must replace the more usual 50ohm version when taking SWR measurements on 75ohm coaxial feeder.

This antenna is a single-band device but it will provide excellent results on 14MHz. Smaller versions can be made for the higher-frequency bands or longer ones for 7MHz.

## The G8ON antenna

Harold Chadwick, G8ON, first described his long-wire system for DX working on the 1.8MHz band (3.5MHz versions could be made) in the September 1957 issue of the *RSGB Bulletin*, and this was followed by a second article in the same journal in June 1966.

Before examining the G8ON antenna, it may help to take a look at the way L H Thomas, G6QB, approached the problem of setting up a half-wave antenna for the 1.8MHz band in an average-sized garden. Fig 67 shows the G6QB approach (*Short Wave Magazine* June 1962) and his 240ft (73m) wire bends back on itself for the final 50ft. The complete wire length is about 10ft short of a half-wave-length on 1.8MHz, and it has its maximum current (and

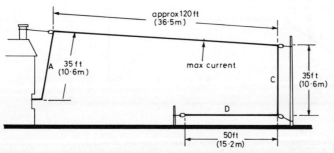

**Fig 67. A 1.8MHz half-wave antenna bent to fit into a 125ft (38m) long garden. This arrangement was described by the late G6QB in 1962. The position of its current maximum means that it is not suitable for long-distance work**

therefore maximum radiation) at a point some 40ft from the far end of the 120ft top section. Being somewhat shorter than a full half-wavelength, the impedance at the antenna feedpoint will be 'manageable' by most ATU units.

This antenna's performance is more suited to short or medium-distance working and it will behave rather like a low dipole. Its vertical section coming down from the 35ft mast will not carry much current, so there will be a restricted ground wave and little power will be radiated at the low angles which are essential for long-distance work.

The original G8ON antenna is shown in Fig 68(a). The wires B, C and D are arranged to be a half-wavelength in total at the intended lowest operating frequency, with the vertical portion C being as long as is possible. This vertical section must be at least 6ft from a metal mast and this preferably should be insulated from the ground.

A short single-wire feeder A is used to ease the matching at the ATU. This is small in terms of wavelength and will have little effect on antenna performance. The two wire lengths B and D are at opposite potential, and are virtually the opposite plates of a capacitor which is shorted by the inductance C. This means that a displacement current will flow in the air dielectric between the horizontal wires.

This explanation was first expressed by G6CJ and was confirmed when an electrostatic screen (made from sev-eral parallel wires) was arranged horizontally between B and D. When unearthed it had little effect, but when earthed the strength of the radiated signals dropped, some-times by as much as 10dB.

The version illustrated in Fig 68(b) may be used in shorter gardens. The length from the end point E to C at the centre of the vertical wire must be a quarter-wavelength on the lowest operating frequency. It is suggested that the wire E should be at least 11ft long. This arrangement tends to reduce the displacement current but it can still be very effective. When used at twice the design frequency (as a full-wave wire) the antenna currents in the horizontal wires will be in phase, and so operate as two half-waves in phase. This will give some gain at right angles to the run of the antenna.

In order to determine the correct inductance of L, it must be adjusted to give the maximum RF current in the vertical wire C. A small lamp or RF current meter can be inserted in the centre of C and the number of turns on the coil L may be varied to give maximum current.

G8ON found that his sandy soil did not reduce the efficiency of this antenna, and in fact other workers have found that a good low-resistance ground is not an advan-tage. This might be because the wire D is only about 6ft above ground. A sandy soil and sub-soil will mean that the true earth will be many feet below the surface, and so the operation of the antenna will be enhanced.

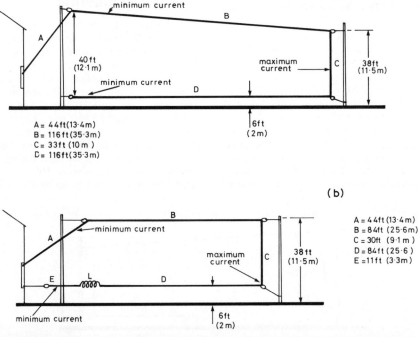

Fig 68. (a) The original G8ON 1.8MHz antenna. The maximum current will be several feet up from the bottom of the vertical section. (b) This later (1966) version of the G8ON antenna uses a loading coil 11ft from the end of the wire. The electrical centre and the point of maximum current is at the foot of the vertical section. Earthing the wire here and removing D, E and L would in no way alter the performance of this antenna!

# Quarter-wave slopers

The quarter-wave or 'half-sloper' antenna is really half of an inverted-V dipole. A quarter-wave antenna is normally arranged to be bottom-fed, which means that the maximum radiation is at the base of the antenna. By inverting the feedpoint of a quarter-wave (see Fig 69(a)) the current maximum and therefore the greatest radiation will be at an elevated position.

The 'ground' against which slopers are fed is usually the metallic mass of the support tower, although a good low-resistance ground at the foot of the tower remains important. The slope angle L (Figs 69(a) and (b)) of a half-sloper is usually 45°, and its maximum radiation is in the direction of the wire away from its feedpoint. The gain of a sloper is said to be between 3 and 6dB over a half-wave dipole but this depends upon the quality of the earthing system below. Buried radials or earth mats etc are needed if the best results are to be realised.

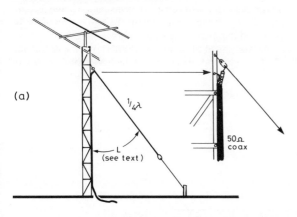

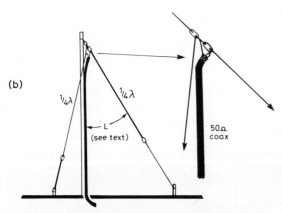

Fig 69. (a) A quarter-wave sloper antenna used with a metal tower. (b) Using a quarter-wavelength of wire which is almost vertical to replace the metal tower. The feed impedance of the sloper depends upon several variables, one being the angle L

The feed impedance of a quarter-wave antenna is normally about 36ohms but a quarter-wave sloper antenna can have a feed impedance which can lie anywhere between 30 and 60ohms! The actual feed impedance depends upon three variables: the length of the wire, the tower or mast height and the enclosed angle between the wire and its support.

If a metal tower is used this becomes a part of the resonant system and there will be a voltage maximum somewhere between the antenna feedpoint and ground. Any other antennas, such as beams, which may be located on top of the support tower will also influence the feed impedance and the antenna's performance. Any guy wires which support the tower must be made non-resonant by breaking them into suitable lengths with insulators.

The version as shown in Fig 69(b) uses a non-metallic support mast, and it has an additional quarter-wavelength of wire (the missing half of the dipole!) dropping vertically or almost vertically down towards the ground. If the feeder SWR is high it can often be brought down to an acceptable figure by changing the wire slope angle. Doing this can result in an SWR of 1.5:1 or better.

An all-band version of the half-sloper cannot be achieved when using 50ohm coaxial feed, so instead an open-wire line (Zepp feed) to a transmatch (ATU) is suggested. The measured bandwidths of half-slopers are about 50kHz at 1.8MHz, 100kHz at 3.6MHz, 200kHz at 7MHz and so on, becoming progressively greater as the frequency increases. A quarter-wave sloper antenna, when designed for operation around 1830kHz (the CW DX sector), can therefore have a rather high SWR when used above 1.9MHz.

On 1.8MHz a support tower more than 110ft (33.5m) high is required, which puts it beyond the reach of most British amateurs. Obtaining planning permission for structures of this height can be very difficult!

# The 'lazy quad'

Although superficially similar to a quad antenna in that it is a full-wave loop, the 'lazy quad' is in fact a close relative of the Kraus 'W8JK beam'. Unlike a quad it is held in a horizontal plane and Fig 70 shows a plan view of the antenna. The centre of the loop is broken by an insulator so it therefore becomes two half-waves, each of which is end-fed. The feed impedance is very high (about 9000ohms) and a low-impedance feeder cannot connect directly to it. The antenna has its maximum radiation in two directions and it is horizontally polarised. The gain over a half-wave dipole at the same height is 4dB (a power gain of 2.5 times), and a 'lazy quad' designed for the 21 or 29MHz bands would go comfortably into the roof areas of many homes.

## Feed arrangements

The easy way to feed a beam of this type is to use a length of tuned line, either of the open-wire variety or instead a commercially manufactured 300ohms impedance ribbon. Just a short run of this feeder is needed if the antenna is located indoors; the tuned line can be easily matched to the required 50ohms impedance with an ATU.

Fig 70 shows a quarter-wavelength long matching transformer made with 675ohm impedance open-wire line which will provide a 50ohm balanced output impedance. At this 50ohm point a 1:1 balun must be used to allow the connection of any length of a standard 50ohm coaxial cable. The quarter-wave transformer is a relatively easy way to provide impedance step-down or step-up ratios, and its own impedance can be deduced from the statement:

Quarter-wave transformer impedance =
√(Antenna impedance x Feeder impedance)

The 'lazy quad' has an antenna impedance of about 9000ohms, so the impedance of a suitable matching quarter-wave transformer is the square root of 9000 x 50ohms which is approximately 675ohms. The length of the quarter-wave section is calculated from the free-space length of an electrical quarter-wave at the operating frequency, which is then multiplied by the velocity factor of the line to be used. A well-constructed open-wire line will have a

velocity factor of about 0.97 so the length of a suitable matching transformer can be found from:

$$\frac{246 \times 0.97}{f(\text{MHz})} \text{ feet}$$

For operation on 21.1MHz the length of such a quarter-wave matching transformer will be 11ft 3in. (3.42m), and on 29MHz it is 8ft 3in. (2.51m). A 675ohm impedance line is easily made with 18SWG wire, spaced 7in. (178mm).

## Construction and setting up

The length of each half-wave element in this antenna (half of the perimeter) will be 23ft 8in. (7.03m) when cut for operation on 21.2MHz and on 29MHz they will each be 17ft 4in. (5.19 m). The insulation at the four corners is not very important, for these are not points of high impedance or voltage, and for indoor use some nylon cord or even household string will suffice. The insulation at the feedpoint and also at the other ends of the half-waves must be good and the use of glass insulators is suggested.

A 'lazy quad' for use on 29MHz only has 8ft 8in. (2.64m) sides and an outdoor rotatable version (only 90° of

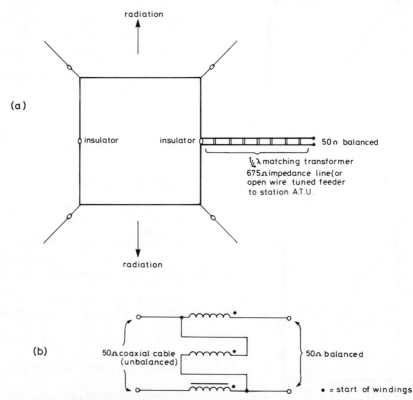

Fig 70. (a) The 'lazy quad' which is a 'split' full-wave loop arranged to lie horizontally. It is ideal for indoor use but its very high feed impedance makes matching to a 50ohm feeder difficult. An open-wire tuned feeder can be used instead of the quarter-wave matching transformer shown. (b) The circuit of a trifilar-wound, 1:1 ratio, unbalanced-to-balanced (or vice-versa) balun. A ferrite rod core is suitable for this balun if moderate power levels are used

turn needed) using bamboo spreaders would be compact. Even on 21MHz the diagonal spreaders could be made with four 8½ft (2.59m) lengths of bamboo or glassfibre rod or tube.

To tune the antenna to resonance, a shorting bar (which can be conveniently made from two 'crocodile' clips) is connected across the end of the 675ohm matching section. If a dip oscillator is then coupled to this shorting bar the resonant frequency of the antenna can be measured. By moving the shorting bar along the line, a point will be found which indicates resonance at the required frequency and then the unwanted wires can be cut and removed together with the temporary shorting bar. If open-wire or ribbon feeder is used as a tuned line to feed this antenna there is no need to tune the antenna to resonance; any length discrepancy will be taken up within the feeder system.

A suitable 1:1 balance-to-unbalance, trifilar-wound balun can be made in a similar fashion to that shown in Fig 39(b). Three, instead of two, lengths of enamelled wire must be bound tightly together as suggested by Moxon before winding them on to the ferrite-rod core. Fig 70(b) shows the connections to the balun windings; any length of 50ohm coaxial cable can be connected to the 'unbalanced' end of the balun. The SWR should be better than 1.5:1 over a bandwidth of 400kHz centred upon the design frequency of 21.2MHz. This bandwidth will be greater on 29MHz.

## The bi-square antenna

This is another loop antenna (Fig 71(a)) but, although similar in appearance to the 'lazy quad', is instead a broadside two-wavelength broken loop set up in the vertical plane. The bi-square antenna has gain in two directions (looking through the loop) of about 4dB over a half-wave dipole and it only needs a single support. This will be 36ft (11m) high for an antenna designed for 21MHz and only 28ft 8.53m) high for a 29MHz version. The total loop wire length is 2 × 960/f(MHz) feet and the antenna is set up in the form of a diamond.

Each side of the bi-square has equal horizontal and vertical radiation components: see Fig 71(b). The latter cancel and leave four horizontal sources which are in phase. This means that the antenna radiation is horizontally polarised at design frequency but, when used at half-frequency, there will be end-fire directivity and vertical polarisation. At half-frequency the gain reduces to about 2dB and the antenna will not have its normal 1000ohm feed impedance. To get the best results this antenna must be at least 3ft above ground and preferably from 10 to 12ft (3 to 3.66m) above electrical ground.

The bi-square is best fed with a tuned line, but for single-band operation a quarter-wave matching transformer can be employed. A fair match to 75ohms will be achieved if a length of 300ohm ribbon feeder is used as the stub or matching section (the output impedance will actually be in the region of 80-90ohms) and this approach simplifies its preparation. When a 75ohm coaxial feeder is used a 1:1 balun should be connected between this and the quarter-wave transformer.

A 21.2MHz bi-square will have two sides of equal length, each one being 45ft 3in. (13.78m) long, while on

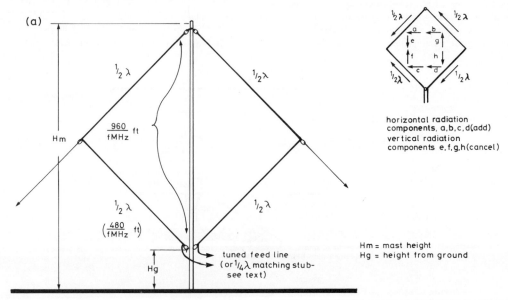

Fig 71. (a) The bi-square antenna gives a useful gain. It is a broadside two-wavelength loop set up in the vertical plane. It can be used at half-frequency with reduced gain and it will then have a different radiation pattern. (b) The instantaneous current distribution along a bi-square antenna. The horizontal radiation components add and the vertical components cancel. Radiation is therefore horizontally polarised and is in two directions, ie looking through the loop

29MHz they reduce to 33ft 2in. (10.1m). If the slotted Bofa 300ohm ribbon is used to make up the matching transformer, it will be either 10ft (3.04m) long for 21MHz or 7ft 4in. (2.14m) for 29MHz. The older unslotted and less-satisfactory type of ribbon feeder has a different velocity factor (actually 0.8) and it will have to be cut to shorter lengths. Good insulation is needed at all the four corners of a bi-square, for these points are at high impedance.

## A switched indoor beam

This antenna is a close-spaced, two-element Yagi beam using a folded-dipole driven element and a parasitic wire which may be switched to operate as either a director or a reflector. It is designed for use on 29MHz and would fit into an average-sized loft or roof space. The author built and used a similar antenna during the peak of solar cycle 19 back in 1957 when the 28MHz band was 'wide open' on most days. Then, for the first time, the author's 'CQ' calls from a 50W AM transmitter stimulated many 'pile-ups' of North American and Australian stations!

A pair of 4ft long (1.21m) bamboo or dowel rods provide the necessary spacing S (see Fig 72). The driven element uses a 14ft (4.26m) length of slotted 300ohm ribbon as a folded dipole A which has additional 1ft (30cm) wires connected at its ends B.

The parasitic element D is 14ft 10in. (4.28m) long and it performs as a director. An additional 2½ft (0.73m) of wire C can be added to D to make this element operate as a reflector. The switching is done with a mercury tilt switch which can be manually operated via a length of thin cord. This cord could progress through the roof space and eventually reach the operating position.

The author's shack is conveniently located just below the roof space in his house and several small holes in the ceiling allow the easy passage of coaxial cables, wires and cords! If it proves impossible to run a cord down to the operating position a solenoid can be employed to 'pull' the mercury switch cord.

Mercury switches can be purchased second-hand from surplus stores and market stalls, and they should not be expensive. It is understood that such switches are commonly used in some items of domestic machinery such as washing machines. Fig 72(b) shows a typical mercury tilt switch, but there are many other different patterns available. The wires which connect to the switch contacts must be included in the lengths of C and D.

## Other points

The feed impedance of a half-wave dipole which has a single close-spaced parasitic element is about 20ohms. However, if a folded dipole is used as a driven element this impedance is raised four times to 80ohms, a value quite suitable for the direct connection of a 75ohm feeder.

The purist may wish to install a 1:1 balun at the feedpoint, but the author has never found this to be necessary when feeding any dipole which has only a short run of coaxial cable feeder. Long feeders may sometimes introduce a 'squint' into the pattern of the dipole or beam, however.

An indoor beam will certainly not need more than a half-wavelength of feeder when the operating position is above the ground floor. The usual and commonsense practice of keeping any indoor antenna wires away from water tanks etc must be followed when locating this beam antenna in a roof space. A theoretical gain of up to 5.5dB over a half-wave dipole is possible with this antenna, but indoor use will certainly reduce this figure by a decibel or so.

If blessed with a 'king-sized' roof space, a 21MHz

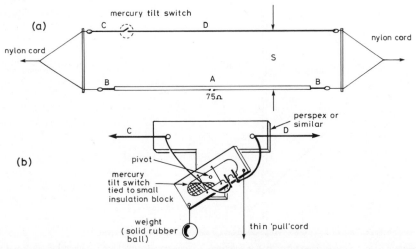

Fig 72. (a) A simple indoor two-element Yagi beam which uses a mercury tilt switch to change the length of the parasitic element. (b) How the mercury switch can be arranged to tilt when a thin cord is pulled. A small counterweight will return the switch to its original position

version of this can be tried and its dimensions (centre frequency of 21.2MHz) will be:

A = 19ft 2in. (5.84m)   B = 1ft 6in. (0.45m)
C = 2ft 8in. (0.78m)    D = 20ft 10in. (6.12m)
S = 6ft 6in. (1.84m)

## The 'jumbo-jay' vertical

The J-antenna has a long history, which certainly goes back to the mid-'thirties, and it is normally used for VHF work (50MHz and up). It is a full-sized half-wave vertical radiator which has at its base a quarter-wave matching section. A half-wave vertical antenna is a vertically polarised and omnidirectional free-space radiator which can give excellent low-angle radiation. Its efficiency is 50% greater than a quarter-wave vertical ground-plane antenna and it does not need radials or any special earthing arrangements.

Centre-feeding a vertical half-wave antenna will mean that its low-impedance feeder would normally come away horizontally from the dipole centre for some considerable distance before dropping to ground. If this is not done, the radiator will be unbalanced and its feed impedance will not remain close to the usual 75ohms. The feed problems are overcome by using a quarter-wave matching section at the foot of the half-wave.

## Construction

F C Judd, G2BCX, used the J-antenna as a starting point for his 'Slim Jim' vertical antenna, which has a half-wave folded-dipole element bottom-fed via a quarter-wave matching section. This feed arrangement is identical to that of the J-antenna. By using tubing for the VHF 'Slim Jim' the antenna becomes sturdy and can be self-supporting.

On the other hand, the 'jumbo-jay' as shown (see Fig 73) is designed for use on 29MHz and is made with wire. The alternative 'Slim Jim' design can also be used on the 28-30MHz band; a scaled-up version (using wire elements) was built and tested, and found to be excellent by the author. It had little advantage over the basic 'J' system and was also a little more costly to build.

A non-metallic support (or suspension from a horizontal rope or wire) which is at least 30ft (9.1m) high is needed to hold up the 'J', and a 16ft 1in. vertical half-wave length of wire is dropped down from the highest point. The matching section is arranged to lie along a strip of marine plywood (or similar weatherproof material) so that the base of the 'J' is about 6ft above the ground, low enough to allow an easy adjustment of the matching when setting up the antenna.

The insulation must be good at the top of the antenna and also where the lower end of the radiator joins the matching section. The latter is a non-radiating, end-shorted, quarter-wave length of twin feeder which has a wire spacing of

about 3in. Ceramic or other suitable stand-off insulators can be fixed to the plywood strip to support this matching section.

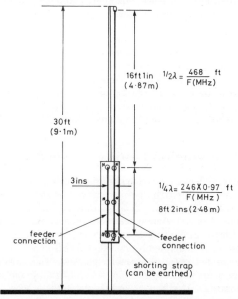

Fig 73. The 'jumbo-jay' vertical antenna is a 28MHz version of a popular vhf antenna. It is a half-wave radiator which is matched to a low-impedance coaxial feeder by using a quarter-wave stub

## Matching

The 'jumbo-jay' must be tuned to resonance before the 50ohm coaxial feeder is connected. A small shorting bar made with a pair of 'crocodile' clips with a half-turn coil of wire between them can be easily loosely coupled to a dip oscillator. Starting at the lower end of the matching section, the shorting bar should be moved up slowly while continually checking the resonant frequency with the DO. When the correct setting is found (29MHz), the temporary shorting bar can be replaced with a wire soldered across the matching section and the unwanted remainder cut away.

Using low power (at the resonant frequency), and with an SWR meter in the feedline, the coaxial feeder is connected temporarily (again by using 'croc' clips) across the lower end of the matching section, a few inches above the shorting bar. The tapping points must be adjusted to get the lowest SWR reading and then permanent soldered connections may be made. A perfect match showing an SWR reading of 1:1 may not be attainable but if this reading is 1.5:1 or better the losses will be very small. The centre of the shorting bar can be earthed and this will prevent any build-up of static charge on the antenna.

## Performance

The bandwidth of the 'jumbo-jay' will not be so wide as that of a similar-sized antenna constructed with tubing, but it will allow operating over the FM and upper SSB segments of the 28MHz band. With just 4W output power, the author's version of this antenna could put down S9 ground-wave signals at distances of 30-40 miles. Unfortunately the antenna's long-distance capabilities could not be tried, for when it was being tested there were no DX openings.

The commonly found screening factors at some locations (caused by buildings, trees and rising ground etc) will distort the theoretical all-round radiation pattern of this antenna, particularly as its high-current section is only about 22ft (6.7m) from the ground. When the antenna has been set up and matched it could of course be relocated in a more elevated position (ie on top of a building) and then its performance would be greatly enhanced.

## Small shielded loops for receiving

The reception of weak signals on the 1.8MHz band is often seriously hampered by the interference of man-made noise. We are all now paying dearly for the convenience of the many electrical equipments and gadgets that seem to proliferate as our living standards improve. The author's location, although semi-rural, is not free from a high radio background noise of variable intensity. This emanates from many sources, including central heating and other thermostats, food mixers and electric cattle fences.

A good vertical antenna system for 1.8MHz seems particularly adept at picking up such unwanted noise from quite distant sources, so the author determined to experiment with small screened loop antennas. Their design was based upon that described in *QST* for March 1974 by Doug DeMaw, W1FB, which has since also been published in *The ARRL Antenna Handbook* (14th edn).

The actual geometry of the loop (see Fig 74) is not critical but an important criterion is the area it encloses. The larger this area is for a given length of loop, the greater will be the received signal strength. A true circle (a) is the best configuration, followed closely by the octagon (b).

The remaining shapes (c), (d), (e) and (f) have diminishing levels of effectiveness. The square at (c) is perhaps the easiest to make and support and it is almost as effective as the ideal circular shape. An equilateral triangular form was tried by the author but this was noticeably inferior to a square loop.

Small loops have a bidirectional figure-of-eight pick-up pattern (see Fig 75(b)) which is similar to that of a half-wave dipole, the difference being that the maximum signal strength is received from directions that are in line with the axis of the loop and *not* at right angles to it. This is the opposite of the patterns of the quad and delta-loop antennas. At right angles to the axis of the small loop there is a very sharp and deep null which may be as much as 30dB down when compared with the optimum pick-up direction.

The term 'small' when discussing loop antennas means any loop which has a conductor length of no more than 0.1 wavelength. The shielded loop to be described is only 0.036 wavelength long, which at 1.8MHz is 20ft (6m), and this length is just about the maximum possible dimension if it is to be tuned to resonance. The distributed capacitance of the coaxial cable used to make up the shielded loop is too great when a larger loop is constructed.

### Constructional details

A representational and not-to-scale drawing of a typical

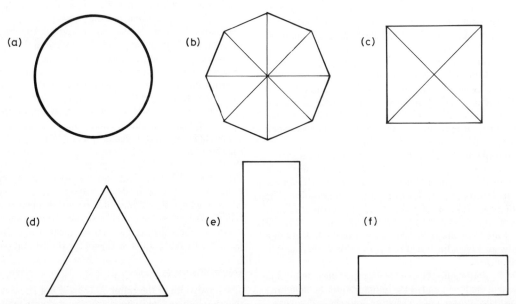

Fig 74. The shapes shown in (a) to (f) are all possible arrangements for single-turn shielded-loop antennas. They all have the same total side length but different enclosed areas. The greater the enclosed area the more effective a receiving loop antenna will become

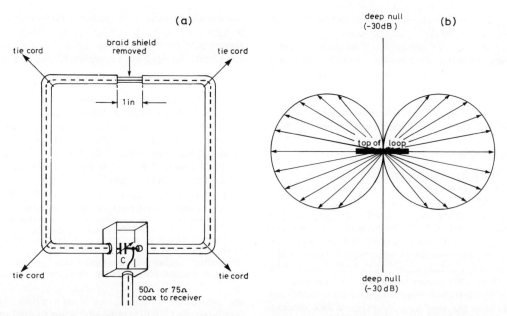

Fig 75. (a) Constructional details of a 20ft shielded loop for use on 1.8MHz. The coaxial cable used for the loop must have a low self-capacitance or it will prove impossible to resonate. (b) The directional characteristics of the loop antenna. The deep nulls at right angles to the loop are an important characteristic

1.8MHz shielded loop is shown in Fig 75(a). A 20ft length of either 75 or 50ohm coaxial cable is arranged in a square and has its ends coming to a small metal box which contains a variable capacitor C. The coaxial cable used *must* be of the low self-capacitance variety. Tests made upon several types of cable revealed a very wide range of capacitance values.

The worst example tested was a piece of cheap 50ohm cable sold for CB use; this had a measured capacitance of 60pF per foot! It would have presented a total capacitance of 1200pF when used as a 20ft loop antenna, and the system would have proved impossible to resonate on 1.8MHz.

The best cable was a semi-air dielectric type which measured out at approximately 16pF per foot. Some UK UR series of coaxial cables which have low self-capacitances are UR57 (20.6pF/ft) and UR63 or UR85 which both only measure out at 14pF/ft. The better cables use a helical membrane dielectric. The American type RF-11/U foam (16.9pF/ft) and the non-foam variety of this cable (20.6pF/ft) can be used successfully when making a shielded loop antenna. In general, the air or semi-air spaced and the foam dielectric types are the most useful cables for making receiving loops.

A 20ft loop will have an inductance of about 15µH so a total capacitance of about 500pF is needed to tune it to resonance. Such a loop, made with a good coaxial cable having a capacitance of 20pF/ft, will therefore need approximately another 100pF to resonate it on 1.8MHz. The author's loops, which are made with a very-low capacitance cable, need an additional 180pF for tuning.

Many of the cables in common amateur use have a capacitance of about 30pF/ft and their total capacitance, if they were used to make a 20ft loop, would be 600pF, which is too great to allow resonance at 1.8MHz. The capacitor C (Fig 75(a)) can be a standard-type broadcast-receiver tuning component, having a maximum value of about 360pF.

The loop must have about 1in. (25mm) of the coaxial braid removed at its centre point or there will be no signal pick-up at all! Doing this will let the braid behave as a Faraday screen, and makes the loop perform as a 'magnetic antenna' which is not influenced by the proximity of metal or other objects. (In Germany shielded loops are known as 'magnetic antennas').

The shielding greatly reduces the pick-up of electrostatic pulses of radiation and this factor gives the antenna a superb signal/noise figure. A screened loop can be used successfully either indoors or quite close to the ground with little adverse effect upon its performance. Even a location close to metal heating pipes, metal water tanks and electric wiring etc in a house loft is quite usable.

## Tuning the loop

An initial check that the author's first shielded loop would resonate on the 1.8MHz band merely involved laying it out on the floor of a room adjoining the radio shack, connecting it to the station receiver (tuned to 1.85MHz) and then adjusting C for the maximum background noise. Then the loop was taken up into the loft and tied with string to four points that made it into a vertical square. An extension loudspeaker was taken up into the loft and the loop was once more tuned for the highest noise output.

Some time later the author put up an identical loop at right angles to, and about 20ft away, from the first model. This meant that by switching the loops it became possible to make use of their nulls to advantage in reducing unwanted QRM. The low inherent $Q$ of this type of loop means that if it is initially tuned to mid-band it will operate well over the whole band, ie between 1810 and 2000kHz.

## Performance

Some writers have stated that shielded loops do not exhibit directional properties when receiving long-distance signals, but this is certainly untrue. The author's loops demonstrate definite signal-strength advantage in their preferred directions when receiving amateur signals from outside Europe. When listening on the 'wrong' loop distant stations are often too weak to copy properly, but switching antennas will bring these signals up by 10dB or more. In the winter many USA amateur signals can be received at a genuine S9 plus 10 or more decibels on the 'northwest' loop, while on the other one they are below S9 and have to compete with many very strong European signals.

A shielded loop used on 1.8MHz with a Ten-Tec Corsair transceiver and no additional RF amplification reduced the background noise level by as much as 35dB! Tests over a long period have shown that the loops can often give a signal-to-noise advantage of between 15 and 20dB.

## A low-noise broadband amplifier

Although the signal-to-noise advantage of the shielded loops over the author's main vertical 1.8MHz antenna was considerable, it was decided that some RF amplification of the signals from the loops would be an advantage. A low-noise broadband amplifier was built, and the circuit of this is shown in Fig 76. It is based upon a design by Wes Hayward, W7ZOI, which appeared in *The ARRL Antenna Book*, and is widely used for amateur RF and IF amplification stages.

The amplifier is inherently stable for two reasons: the degenerative feedback in the emitter and also the negative feedback in the base circuit. A broadband 4:1 transformer in the collector circuit is used to step the output impedance down to about 50ohms, and there is additionally a 50ohm characteristic at the input to the BSX20. (The transistor used in the original W7ZOI design is a 2N5179 which is difficult to obtain in Britain.)

The back-to-back diodes in the output offer some protection to the transistor should the transmitter output be accidentally routed through the amplifier, and those across the input limit the RF pickup voltage from the loop which is always present when transmitting on the main antenna. The bifilar wound 4:1 transformer can be made with an Amidon toroidal core type FT-50-61 or similar. A short length of 1/4in. diameter ferrite rod can also be used to make the transformer if a toroid core is not available. The coil windings are 12 bifilar turns (the wires twisted together before winding) using 24 or 26SWG enamelled wire.

This amplifier will provide about 15dB of gain with virtually no degradation of the signal-to-noise ratio. It matches to the loop feeder coaxial cable and the receiver, and also makes the copy of weak signals more comfortable.

## Outdoor use

If a shielded-loop antenna is to be used out of doors it is suggested that it is housed inside a length of plastic tube. Three 2m lengths of 3/4in. diameter plastic 'waste pipe' can be joined to provide a suitable circular hoop, which after

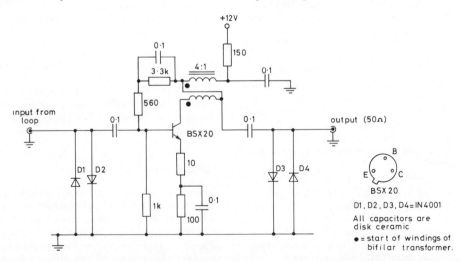

**Fig 76. The circuit for a low-noise wideband preamp for loop antennas. It may be situated close to the loop or at the receiver end of the feeder**

weatherproofing etc may be suspended or even arranged to be rotated. The base of the loop only has to be 3 to 4ft from the ground so a very short support pole will suffice.

Half-sized loops of this type can also be made to resonate on 3.6MHz where they would be useful for the reception of DX signals. On this band they would be advantageous in nulling out many of the strong European stations.

## Beverage antennas

The Beverage antenna is also called a 'wave antenna'; in its simplest form it is just a horizontal wire ranging in length from 0.5 to 10 wavelengths and supported at about 10ft above the ground. It was first developed by H H Beverage as a long-wave receiving antenna at Riverhead, USA. His original antenna was in fact 10 miles long and it was positioned over dry sandy soil.

If the received waves to such an antenna were to be exactly vertically polarised when they reached the horizontal wire, no RF voltages or currents would be built up along its length. The dielectric properties of the dry soil at Riverhead gave the wave a considerable forward tilt, so that a horizontal component developed which induced voltages along the wire.

The signals Beverage was receiving were essentially vertically polarised ground waves, but amateurs receive sky waves which have mixed polarisation and usually arrive at the antenna at many different wave angles. This means that an amateur Beverage antenna is not so dependent upon having the dielectric properties of the dry soil beneath the original Beverage antenna. However, it is still found that a Beverage is more effective when it is over low-conductivity soils. The wave tilting involved when receiving vertically polarised ground waves gave rise to the term 'wave aerial'.

An early article in *Wireless World*, not long after the first world war, was written by a signals officer who had served in the Tank Corps. He wrote that all the 30ft masts supplied as antenna supports were rapidly shot to pieces within minutes of their erection when operations were taking place near the front line, so he and his contemporaries often used long wires laid out on the ground. These wires were very good when they pointed towards desired stations but they did not pick up any signals from other directions.

Such long wires at ground level were an early form of wave antenna. Wave antennas are now only used for reception, although 50 years ago versions were developed to work at HF (above 2MHz) for transmitting.

In Fig 77 a Beverage antenna is shown running along a line from east to west and a transmitting station is assumed to lie to the east. The travelling wave from this station, when it reaches the antenna wire, will move along it from east to west and induce currents in the wire which then travel in both directions.

The current travelling east moves against the motion of the wave and it reduces to almost zero when the wire is one wavelength or more long. The currents travelling west, however, travel at almost the velocity of light and will therefore move along with the wave. These currents moving west all add up in phase at the west end and produce a strong signal there.

If the eastern end of the wire was either grounded or open-circuit, the induced currents generated by signals from either the east or the west would be reflected back to the western end of the wire. The antenna would then become bidirectional.

The terminating resistor R absorbs the RF energy reaching the far end of the wire, so preventing any reflection, and gives the antenna a unidirectional property. The resistance of R must match the surge impedance of the antenna wire, and this depends upon both the thickness of the wire and its height above the ground. It usually has a resistance of between 200 and 400ohms and must be non-inductive.

A Beverage works best over a poor earth but the earthing to which the terminating resistor connects must be very good. Several radials or buried wires should be used as is shown in Fig 77. A simple transformer steps down the high impedance of the Beverage wire to a low value which is

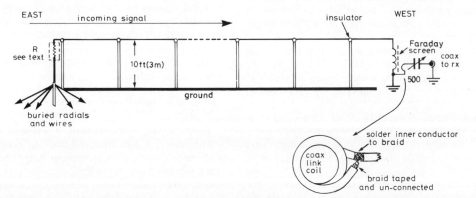

Fig 77. The terminated Beverage receiving antenna for use on 1.8 or 3.5MHz. The Faraday screen is made from one or two turns of coaxial cable. To be effective a Beverage antenna must be at least one wavelength long

suitable for a match to a coaxial cable. The secondary of this transformer should be electrostatically shielded from the primary winding, and this is easily managed by the use of a two (or more) turn link winding made at the end of the coaxial cable. This is arranged as a Faraday screen (see Fig 77).

## Length and performance

One wavelength must be regarded as a minimum length for an effective Beverage antenna, and on 1.8MHz this will be about 500ft. Single-wire Beverages have an optimum length of from one to three wavelengths and an optimum height above ground of 10 to 20ft.

A terminated Beverage gives strong signals from stations that are located away from its far end, and will have almost no pick-up at all from its sides. It also has a very low noise level with a considerable attenuation of atmospheric noise. There is no other antenna type its equal for the reception of DX on the 1.8MHz amateur band. Loops are good but are really not in the same class as a Beverage wave antenna.

Unfortunately point-to-point working is not often very important to amateurs, for they generally wish to contact different distant stations which may be located anywhere in the world – several Beverage antennas would be needed to do this. Few amateur operators are fortunate enough to have the ground which is needed to set up one or more Beverage antennas, so most workers on the LF bands must use other low-noise systems. Further reading and much practical information on Beverage antennas can be found in the ARRL book *Low-band DXing* by John Devoldere, ON4UN.

## The 'fishtail' five-bander

Although not a wire antenna, the broadband 'fishtail' is not made with tubing and it is a cheap and easy way to set up an indoor radiator for five HF bands. Kitchen foil is easily obtained at most food stores and it comes in a wide range of widths and lengths. The widest available foil should be used to make this antenna and this will usually be between 30cm and 45cm in width. Rolls of foil up to 20m in length can now be obtained.

By cutting a 'V'-shaped notch in the ends of a pair of foil rolls, a broadband dipole can be made which will operate with a low SWR on the bands between 14 and 30MHz. The cut foil extends from the feedpoint (see Fig 78(a)) in two directions to a range of lengths between 8ft and 14ft 7in.

Although the total extent of the dipole is a few feet short of the correct half-wavelength on 14MHz, it will still work quite well on that band, for a 'fat' antenna will resonate when it is considerably shorter than the 'normal' wire length.

A simple 75 or 50ohm coaxial feed can be used without a balun, but some care must be taken when connecting the feeder to the inner ends of the antenna. Soldering is

impossible so instead several small plated 'crocodile' clips, which are wired in parallel, can be used to afford reliable and low-resistance connections. The foil should be folded back on itself for a few inches to make a strong pad where the clips can connect. The weight of the coaxial cable must be taken off by tying it to a convenient roof timber.

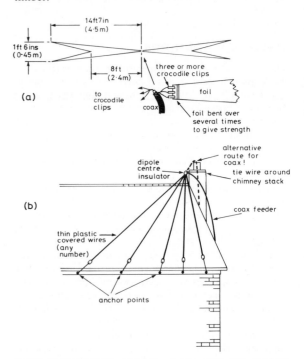

Fig 78. (a) Details of the 'fishtail' antenna, which is made from aluminium foil. Its broadband characteristics mean that it can be used over five of the amateur HF bands. 'Crocodile' clips are used to connect the feeder to the inner ends of the antenna. (b) How a broadband dipole can be arranged over a house roof. This antenna will hardly be noticeable if plastic-covered wire which is coloured to match the roofing materials is employed

This antenna can be arranged to run the full length of a roof and actually be stuck to the underside of the slates. Most modern homes now have their roof slates or tiles covered with felt or hardboard in the roof space and this can make it easier to affix the antenna.

The author described a similar antenna in an article which was published a few years ago, and soon afterwards had a letter from an enthusiastic user who said he had 'raised' a ZL station on 14MHz SSB after his first 'CQ' call!

Before setting up the 'fishtail' it is best to experiment with a variety of adhesives and discover which one will work satisfactorily with aluminium foil. Where the antenna narrows towards the feedpoint the foil can be bent under. This will add to the strength and it may even be possible at some locations to fix the antenna in position with drawing pins (thumb tacks).

## A broadband rooftop dipole

There are now an increasing number of restrictive cove-
nants which present-day house owners are obliged to
observe. These can apply to such things as a restriction on
hanging out lines of washing, a ban on certain pets and,
more seriously for a radio amateur, a total embargo on
radio masts or poles. In some places nothing can be erected
that is higher than the roof ridge or the chimney stack, and
even TV antennas fall within these draconian regulations.

The multi-wire dipole shown in Fig 78(b) is similar to
the multiband antenna described in Chapter 1 and illustrated there in Fig 9. However, its centre is attached to the
brickwork of a chimney stack and it has several wires
descending and lying just above the roof to fixing points
along or near the rain gutter.

If the shortest pair of these dipole wires are cut to make
a half-wavelength on the highest frequency band to be used
and the longest wires are made resonant on a lower band,
several intermediate wires can also be set out. The more of
these there are, the wider will be the bandwidth of the
dipole. The antenna wires can be plastic covered and their
colour chosen to match the roof colour. They can use quite
small unobtrusive insulators.

A simple coaxial feeder (not the heavy ¹/₂in. diameter
variety) of 75 or 50ohms impedance can be dropped down
the end wall of the house, or instead taken into the
chimney! This can only be done if the chimney is not in use,
and fortunately these days most chimneys become redundant as other less-primitive heating systems are used to
replace fires. If the cable is taken down the chimney some
precautions against the trickle downwards of rainwater
must be taken.

This antenna system has its high RF current point as
high as is possible without the use of a mast and it will
certainly out-perform most indoor antennas. The old timers' adage that "A foot of wire outdoors out-performs ten
feet inside" is often true.

## Subsurface (underground) antennas

Try not to laugh, for this section is quite serious! Underground antennas have a long history and were even being
considered as early as 1912. They were tried during the
'twenties (see *Amateur Wireless* September 1922), and at
the present time there is much secret research going on all
over the world in attempts to perfect the ultimate 'invisible' antenna. A nuclear blast would destroy just about
every type of above-ground antenna system so a buried
radiator seems to be the only 'hardened' system which
might survive such a catastrophe. So much for the dishes
festooning those tall antenna towers which seem to domi-
nate most high points in our countryside!

The author was rash enough to write an imaginary or
'spoof' April Fool article on the subject of underground
antennas a few years ago. To his surprise, he received a

letter from Richard Silberstein, W0YBF, who suggested
that several of his assumptions were actually correct!
W0YBF has been involved in research on this subject for
many years and some of his conclusions can be found in the
*ARRL Antenna Compendium* Vol 1 ('Subsurface Antennas
and the Amateur').

The RSGB *T & R Bulletin* published in February 1927
an article by C H Targett, G6PG, who was one of the
foremost British underground antenna enthusiasts. His
antenna was a 60ft rubber-insulated wire about 2¹/₂ft below
the ground. The wire was supported on small posts which
had insulators at their tops and was enclosed within an
arrangement of discarded semi-circular 'pan tiles' before
the trench soil was replaced (see Fig 79).

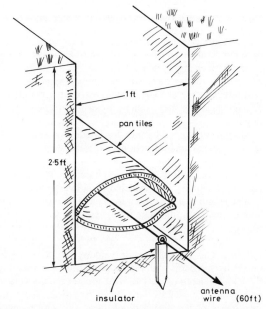

Fig 79. An early attempt at making an underground antenna. This
was fully described in the RSGB *T & R Bulletin* in 1927 by G6PG

G6PG discovered that this antenna, which was end-fed
and reached the surface through a length of rubber hose,
reduced QRN almost to zero and allowed the reception of
shortwave signals almost as efficiently as did his elevated
wire. His 8W input transmitter was fed to the underground
wire on wavelengths of 150-200m, 90m and 45m, and he
had many contacts up to a range of about 1000 miles.

A significant fact was that this antenna was very directional: all the stations worked lay within an angle of 30°
from its far end. This is similar in fact to the performance
of a terminated non-resonant wire antenna or a Beverage
antenna.

Fig 80 shows an underground antenna which was used
for tests by W0YBF in 1965. With this he received signals
from WWV on 5MHz when that station was located at
Beltsville, Maryland. All soils differ in their attenuation of

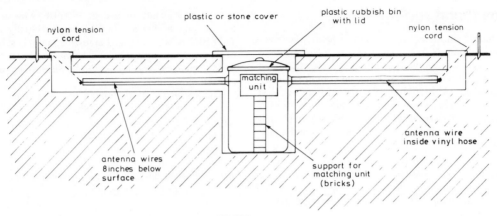

Fig 80. A 1965 subsurface antenna designed and tested by W0YBF

signals, and the term 'skin depth' is used when working with buried antennas. The skin depth of a soil is the depth at which the electric field is attenuated by 1 neper (8.68dB). Fortunately soil is a dielectric and a very poor conductor at radio frequencies; at 5MHz a radio wave when penetrating an average soil is attenuated by about 1.87dB per metre of depth.

Do not get excited by this, for in addition there is also some reflection and refraction loss: Silberstein's antenna, which was 8in. (20cm) below the ground, had an average signal loss of 16dB when it was compared with a standard half-wave dipole at 0.3 wavelength above ground.

The length of a resonant wire when it is buried is much shorter than a similar wire that is surrounded by air as a dielectric. The 5MHz antenna used by W0YBF had to be shortened from the normal 93.6ft (28.53m) to only 46.6ft (14.2m). This means that the velocity factor was influenced by the earth's dielectric properties and measured 47.4% at a depth of only 8in. At greater depths velocity factors as low as 25% have been recorded. The centre impedance of the buried dipole was low, and a special matching unit was used at the antenna centre to allow a good match to a standard coaxial feeder.

W0YBF admits that in 1965 he did not have very good insulating materials (they did not exist), and suggests that experimenters should try dipoles in wide-diameter PVC water pipes or plastic sewer pipes. Some internal supports for the wires would be needed if these were to be used. He resonated his antenna with the help of a dip oscillator when the lid of his waste bin at the dipole centre was removed!

The author's ideas for a similar antenna are shown in Fig 81. Here a wide plastic pipe is suggested, with a large quantity of expanded polystyrene 'chips' surrounding the pipe (this material is used for packaging) and just a thin top

layer of soil; say, 2 to 3in. (5 to 7cm). The centre of the antenna could be made accessible for tuning and the coaxial feed would of course be buried. The underground antenna is perhaps the last ray of hope for the amateur in a zone where the rule is 'strictly no antennas'.

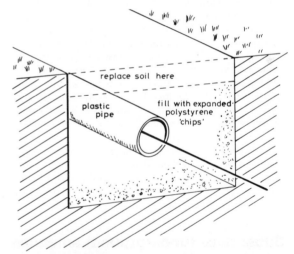

Fig 81. How an underground antenna system might be designed using modern materials

If only buried a few inches and with a lot of polystyrene chips, the attenuation on transmit and receive could be as little as 10dB as compared with a properly set up half-wave dipole. This power loss could be easily made up at the transmitter end of the system. On receive, although signals would be down, there would be very little external noise pick-up.

# Chapter 7

# Antenna matching systems

Designing and putting up an antenna system is not often too difficult, but when this has been completed there remains the problem of how to effectively transfer the output power from the transmitter into that antenna. Quite often much of the output power from amateur stations is lost, either as heat or as undesired radiation from a length of antenna feedline.

Most of the antennas described in this book should be used in conjunction with an antenna tuning (or matching) unit. This is more generally known as an 'ATU', 'ASTU' (antenna system tuning unit) or 'AMU' (antenna matching unit).

The author always employs such a matching device between antenna and transmitter, for it helps in several ways. It limits the strengths of unwanted off-frequency signal inputs to the station receiver, minimises harmonic radiation and presents the correct load impedance to the equipment. Much modern equipment will only operate properly when it 'looks' into a 50ohm load, and it will either suffer damage or instead automatically reduce power if this condition is not fulfilled.

There are two matching problems to be solved: the matching of the antenna to the feeder and the match of this feeder to the transmitter or transceiver.

## Antenna-to-feeder matching

This problem has been discussed at some length in the various descriptions of antennas in earlier chapters, but some additional reminders or 'pointers' are in order. When the antenna feed impedance is about 75ohms (as is the case when using a typical half-wave dipole), a coaxial or twin-wire feeder having this impedance can be used. If instead a 50ohm impedance coaxial feeder is used with a 75ohm impedance antenna, the additional losses induced by the mis-match will not be very great. Other antenna systems which have a higher feed impedance can be fed with good-quality 300ohm ribbon feeder or an open-wire line that has been specifically made to suit the antenna impedance.

However, there are some 'awkward' impedances, in particular those which are low and lie between 20 and 50ohms and for which standard coaxial and other cables are not readily available. One way around this matching problem is to use two equal lengths of a standard coaxial cable in parallel. This technique was described in Chapter 5 (also see Fig 53), where two lengths of 75ohm coaxial cable are paralleled to provide a 37ohm impedance which will match into the base of a typical quarter-wave Marconi antenna.

Two pieces of 50ohm coaxial cable will give an impedance of 25ohms and, by using two equal lengths of 75ohm and 50ohm cable (they must have the same velocity factor) in parallel, an impedance of 30ohms is achieved. Even two lengths of 300ohm ribbon feeder can be taped together and, when connected in parallel, they will provide a 150ohm feeder.

The use of quarter-wave matching transformers has also been previously discussed (see Chapter 6 – the 'lazy quad' antenna), and such transformers can be easily made when the three impedances (antenna, feedline and quarter-wave section) will result in a 'sensible' arrangement. Unfortunately a good match to standard coaxial cables is not always possible when using this scheme, for the characteristic impedance of the transformer may be difficult or almost impossible to devise.

## The half-wavelength line

A little-used technique is that of the 'half-wave line' system. This is based upon the fact that the impedance at any point along a wire or a feedline is repeated at half-wave intervals, the actual line impedance having no influence upon this characteristic. For instance, if a half-wavelength of 300ohm feeder is connected at one end to the centre of a half-wave dipole (75ohms), it will present a 75ohm impedance at its other end.

The use of half-wave feedlines can be impracticable on the lower-frequency bands unless the antenna is located a long way from the operating position, but there is no reason why any excess feeder (if coaxial cable) cannot be coiled or looped in some way. On the higher frequencies such as 28MHz the half-wave line is rather short (12ft of coaxial cable), and it might only be useful when the antenna is indoors or very close to the operating position.

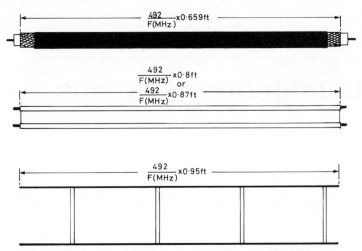

Fig 82. How to calculate the length of a half-wave line by multiplying the free-space half-wave length by a line's velocity factor. The older 300ohm ribbon has a velocity factor of 0.8, whereas the slotted variety has a velocity factor of 0.87

In order to calculate the electrical half-wavelength of a feedline, the free-space half-wavelength of 492/*f*(MHz) feet must be multiplied by the velocity factor of the feeder or cable. Most of the commonly used coaxial cables have a velocity factor of 0,66 (see Fig 82).

When a half-wavelength line is used the antenna feed impedance, even if it has an 'awkward' value, will be translated to the end of the line where it can be matched to the equipment via a suitable ATU. The losses along such

a line will depend upon the characteristics of the coaxial or whatever other type of feeder is used. For most coaxial cables the loss per 100ft (30m) at a frequency of 10MHz will be about 0.6dB. This will rise to about 1.2dB at 30MHz.

Open-wire line and the 300ohm ribbon feeders have much smaller losses than coaxial cable; Fig 83 shows the losses over 300ft of three different types of feeder at 10MHz. It is worth adding that the losses in any coaxial

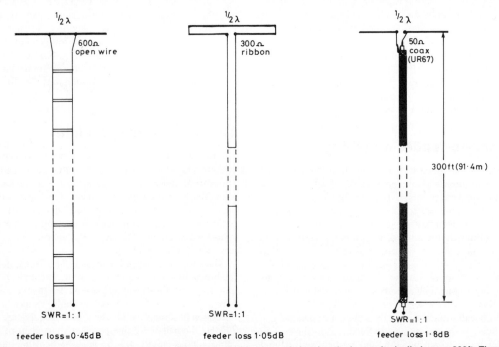

Fig 83. Three half-wave dipoles, each fed with a different type of feeder and showing the losses in decibels over 300ft. These losses assume an SWR of unity. The frequency chosen is 10MHz and losses will be greater at higher frequencies

cables increase when their outer insulating sheaths age and are weathered. If any moisture reaches the copper braid, the resulting oxidation will increase line losses alarmingly.

## Standing-wave ratios

Much nonsense has been written about SWR measurements on feedlines and it is well worth reminding ourselves that such measurements should be kept in perspective. We hear and read of too many SWR 'horror stories'. Not so long ago, when transmitters had valve output stages, very little attention was paid to the SWR (sometimes called the 'VSWR' (voltage standing-wave ratio)) along antenna feedlines.

Transmitters, ATUs and antennas were adjusted to get the greatest power levels into the antenna systems, and any impedance mis-matches, unless they were alarmingly great, were ignored. Such mis-matches meant that the transmitter output valves were not being fully loaded, a condition which did not, however, result in the 'sudden death' that can happen to modern solidstate output stages in similar circumstances.

Most commercial equipments now have built-in safety circuits which reduce the power of their output stages when they are presented with a serious mis-match and in this way prevent the failure of an expensive device. It is this concern for transceiver output stages which has prompted so much attention to the correct matching of transceiver outputs to antenna circuits, and it therefore makes the proper use of an SWR meter or other indicator almost mandatory.

Fig 84 shows a SWR meter connected in the low-impedance feedline from an antenna. The SWR meter must be suited to the impedance of the line that is used; this will normally be 50ohms. If a low-pass filter is needed to limit the radiation of transmitter harmonics it must be placed between the SWR meter and the equipment.

The author finds this arrangement very unsatisfactory, for very few antennas will present a perfect match to the feedline used and therefore there will always be a mis-match between the feeder and the transmitter. This will either bring about a reduction in power output (for safety reasons) or, if special circuitry is not present, cause output-stage overheating.

This arrangement also has no additional tuned circuit between the transmitter and the antenna, so it might allow the radiation of spurious emissions from a typical broad-band output stage. The arrangement shown in Fig 84(a) can only be recommended at best for low-power portable or emergency operations.

Fig 84(b) illustrates a much better way to match the antenna to the transmitter. The ATU is used to match the feeder or the wire antenna impedance to the 50ohm coaxial cable connecting to the equipment (via the SWR meter) and, when correctly adjusted, a near-perfect 50ohm match can be made. The losses introduced by a well-designed

ATU are negligible, although the author has come across some amateur-built ATUs which had losses of 80% and more, and had such high circulating currents that the soldered connections to their coils actually melted!

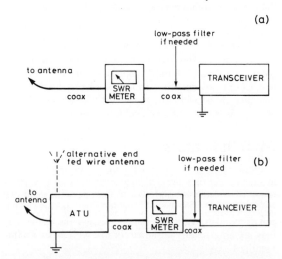

Fig 84. (a) An amateur station which has the transceiver connected to the antenna coaxial feed via an SWR meter. This is not recommended. (b) This arrangement uses a matching unit or ATU between the SWR meter and the antenna. A low-pass filter may be needed if the ATU has high-pass characteristics

An SWR meter measures both the forward voltage (or current) and the reflected voltage (or current) on the feeder. The ratio of the two readings is the SWR and it is indicated by the meter. If a feeder with a characteristic impedance of 50ohms is connected to an antenna load of either 100ohms or 25ohms, there will be a mis-match and the SWR meter will read 2:1. However, the SWR reading does not tell you which of the two possible antenna impedances connects to the feeder. Such a level of mis-match is not really serious unless the losses in the feeder are also large.

There is a multiplication factor of line loss which applies when there is a non-unity SWR on the line. When the SWR becomes 2:1 this factor is 1.25. For example, at 10MHz a 300ft coaxial line which has an inherent loss of 1.8dB (see Fig 82) will have a total line loss of 2.25dB when the SWR becomes 2:1. When an SWR is a poor 3.7:1 the total loss on the 300ft feeder will rise to 3.6dB. An SWR of 3.7:1 will always result in the doubling of the inherent line loss.

Losses of 1dB or so are insignificant. It is a fact that a signal rise or fall of 1dB represents the smallest change that is noticeable by the human ear! Even a 3dB loss (half-power) represents little more than a half of an S-point of signal level.

An SWR of 2:1 will result in a loss of only 1.3dB along 300ft of 300ohm impedance ribbon used as a 'flat' untuned liner. The same length of an open-wire feedline will have a loss of just 0.56dB when the SWR is 2:1! (These values are obtained at an operating frequency of 10MHz.)

A given SWR reading does not, however, give any indication of the reactive components on the feeder, and it will not be accurate unless the load is a pure resistance. Such reactive components can reach the transceiver output stage, and may cause parasitic oscillations and other problems.

This short discussion on SWR and the examples given may help to reassure those who feel anxious when their SWR meters read 2:1, for it is only when the reading rises to 3.5:1 or more that action should be taken to improve the matching. Most modern commercial transceivers reduce power substantially at this level (see page 8).

## SWR indicators

There are very many commercially made SWR meters available, and they are designed to operate over a wide range of frequencies and also to operate with differing power levels. Many designs for the amateur constructor can be found in a variety of handbooks and magazines, but they are all generally designed for use with 50 or 75ohm feedlines.

Fig 85 shows a simple device which will indicate the SWR along a 300ohm ribbon feeder. This is quite an old idea and one that the author has used successfully in the past. A pair of 60mA 'fuse' lamps connected 'back to back' are visual indicators which can provide a rough guide to the SWR on the feedline. If the transmitter output power is adjusted so that the lamp nearest the transmitter glows at normal brilliance, and if the other lamp either does not light or is only dimly illuminated, the SWR will be less than 2:1.

When using low power outputs the pick-up loop may be lengthened to as much as 1ft (30cm), but the author discovered that just a few inches of loop is sufficient when using a power somewhere between 50 and 100W. The SWR on an open-wire line can also be roughly determined by using this method. The loops etc can then be fixed to a strip of insulating material such as Perspex (USA Plexiglass), which has the same width as the wire spacing.

A more detailed explanation of standing-wave ratios may be found in several antenna handbooks, and *The ARRL Antenna Handbook*, which covers this subject at length, can be recommended.

## 'Old-time' antenna matching

More than 50 years ago antenna matching was done directly without the assistance of an ATU. Fig 86(a) shows the typical tuned-anode 'tank' circuit of a single-ended valve transmitter output stage. A single-wire antenna is tapped on to the anode coil via a blocking capacitor Cb. This component was only used to keep the high HT voltage (typically between 500 and 2000V) from the antenna. An antenna current meter of the 'hot wire' or thermo-couple type is in series with the antenna which helps to determine the correct tap point down the coil. Even a voltage-fed antenna (high impedance) will show some antenna current so the meter will always provide a reading.

A typical push-pull output stage is shown in Fig 86(b) and an open-wire twin feeder is tapped at points equidistant from the 'earthy' centre of the coil. A correctly balanced antenna and feeder system will show equal RF currents in the two feeder wires. Such circuits are quite primitive and they would in no way reduce the radiation of transmitter

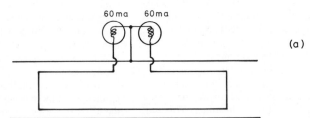

(a)

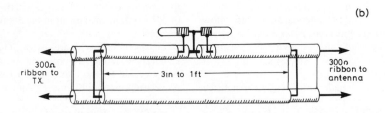

(b)

Fig 85. A simple way to estimate the SWR along a 300ohm feeder. The loop and its two indicator lamps must be taped to the 300ohm ribbon – if just the lamp nearest the transmitter lights the SWR is better than 2:1

harmonics or other spurious emissions. A leaky blocking capacitor might also result in a fatal accident were the antenna or feeder wires then to be touched.

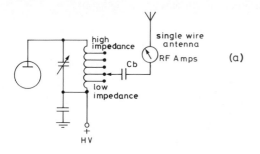

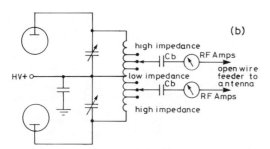

**Fig 86. Single-ended (a) and push-pull (b) valve output stages, showing how antennas were roughly matched many years ago when ATUs were not in general use. The blocking capacitors Cb were vital safety elements as they kept high DC voltages from the antenna wires**

## The parallel-tuned circuit ATU

This is shown in Fig 87(a) and it is similar to the anode 'tank' circuit just described. Although simple, this circuit can be an effective matching unit for single-wire antennas, and it can deal with a wide range of antenna impedances. An ATU should not have a high $Q$, or else energy transfer will be difficult to achieve and the tuning will seem to be very 'sharp'. A $Q$ of between 10 and 12 is ideal, and to attain this the tuning capacitor C and the inductance L1 must each have a reactance of about 500ohms at the operating frequency.

Even when designed within these parameters a very high antenna impedance might still give matching difficulties, and to overcome this the $Q$ may be lowered further by increasing the inductance of L1 and reducing the value of C. A rough 'rule-of-thumb' guide to the correct values of C and L is as follows:

"Let C = 1pF for each metre of wavelength and let L = 0.25µH for each metre of wavelength".

This will mean that an ATU of this type designed for use on 7MHz requires a capacitance at resonance of 40pF together with an inductance of 12µH.

Here are the values suggested for C and L for seven amateur bands when their reactances are 500ohms:

| | | |
|---|---|---|
| 1.8MHz: | C = 180pF | and L = 40µH |
| 3.5MHz: | C = 90pF | and L = 22µH |
| 7MHz: | C = 45pF | and L = 12µH |
| 10MHz: | C = 33pF | and L = 8µH |
| 14MHz: | C = 20pF | and L = 6µH |
| 21MHz: | C = 16pF | and L = 4.5µH |
| 28MHz: | C = 11pF | and L = 3µH |

With transmitter output powers of 100W the tuning capacitor should have a plate spacing of $^1/_{16}$in. (1.5mm) or more.

The link coupling coil L2 has one or two turns of wire which can be set at a variable distance from the tuning coil L1. L2 is wound or arranged to be at the 'earthy' end of L1 and the coaxial cable from L2 should reach the transmitter via an SWR meter. Tuning C, trying the different tap positions for the antenna along L1 and also varying the coupling between L1 and L2 are three adjustments which must be made to secure the lowest SWR reading. For multiband work L1 may use plug-in coils which should preferably be self-supporting and wound with heavy wire or thin copper tube. The same link winding may be used on most bands.

## An ATU for two-wire feeders

A variant of the previous design is shown in Fig 87(b), and it is arranged to match a two-wire feedline. As shown it can accommodate a range of medium to high impedances, and the inductance L1 is tuned by a split-stator variable capacitor. The capacitance across L1 is the effective capacitance of each half of C1 in series. No earth connection to the tuned circuit is shown but the junction of the two sections of the variable capacitor C1 may be earthed. The actual 'earthy' point along the coil will lie somewhere close to its centre and the link winding L2 is arranged over this part of L1.

Fig 87(c) shows another way to match a pair of feeder wires, but in this case it is assumed that the feeder impedance will be low. When this impedance is low a more effective match is made when series tuning is employed, using the two variable capacitors C1a and C1b. Again, no earth connection is required.

### Variable coupling using a capacitor

Both the circuits shown in Figs 87(b) and (c) use link windings L2 which are in series with a variable capacitor C2. This system is much easier to adjust and physically easier to arrange than the variable coupling shown in Fig 87(a).

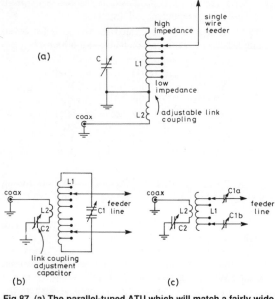

(a)

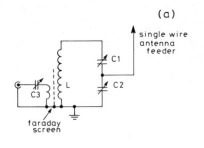

Fig 87. (a) The parallel-tuned ATU which will match a fairly wide range of antenna impedances, but presents problems when the impedance is low. The capacitor C must be a high-voltage working type with wide plate spacing. (b) A parallel-tuned circuit ATU that is suitable for use with tuned feeder lines, particularly when they present a high or medium impedance. A fixed link coil L2 can be used and the degree of coupling may be adjusted by C2. (c) This series-tuned ATU uses a pair of ganged (but not electrically connected) variable capacitors in the feedline and it can match low impedances without difficulty. The coupling link is as described for (b)

The component values for L2 and C2 are found from the lowest frequency to be used in the band of intended operation and also the impedance of the coaxial line to the transceiver or transmitter. The capacitor C2 must resonate with L2 on this frequency, ie it will have a value of 1000pF at 3.5MHz when using 50ohm line if the reactance of L2 also equals this impedance. Some approximate capacitance values for six amateur bands are:

|  |  |  |
|---|---|---|
| 1.8MHz: | C2 = | 1800pF |
| 7MHz: | C2 = | 500pF |
| 10MHz: | C2 = | 350pF |
| 14MHz: | C2 = | 220pF |
| 21MHz: | C2 = | 150pF |
| 28MHz: | C2 = | 130pF |

As the capacitances become rather large at LF, a larger link winding can be used on 3.5 and 1.8MHz, which means a 500pF broadcast-type variable can be used for C2. This capacitor will also be suitable for the other bands.

## The capacitive-tap ATU

Here is another simple circuit which was actually one of the author's favourites for a number of years when he was restricted to a single-wire end-fed antenna. It is similar to

the parallel-tuned circuit as shown in Fig 87(a) but, instead of taps along the coil, two variable capacitors in series are used to provide impedance matching and additionally tune the circuit to resonance (see Fig 88(a)).

(a)

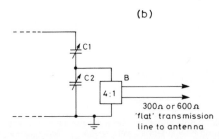

(b)

Fig 88. (a) A capacitive-tap ATU which can give an infinitely variable impedance match to the tuned circuit L, C1/C2. C1 and C2 are separately tuned and the ratio of their capacitances determines the match. The link coil uses a Faraday screen of the type illustrated in Fig 77. (b) A 4:1 balun may be used to present an unbalanced impedance to the capacitive-tap ATU. This can only be done when the transmission line is of the 'flat' untuned type, such as the feed from a folded dipole when employing a 300ohm ribbon feeder. A balun must never be used with tuned lines

If we imagine that the antenna wire is at high impedance, it will match when C1 is large and C2 is small in capacitance. On the other hand, if the impedance of the wire is low, C2 will have to be large and C1 will be small. The normal rules of antenna matching $Q$ will still apply, so a pair of quite high-value capacitors are needed: about 500pF maximum for each. As they are always in series and tune across L their plate spacings can be smaller than would be usual when just a single capacitor is employed.

In Fig 88(b) either a 300ohm or 600ohm 'flat' and non-resonant feeder line can terminate at a balun (either 1:1 or 4:1 ratio) which will present an unbalanced output to the ATU. A balun *must never* be used in this position when a tuned line is employed, for the unwanted reactances can give rise to balun heating and there will be a wasteful loss of power. The balun could even be permanently damaged.

A Faraday shield can be used in the coupling circuit to remove any of the antenna's capacitive effects from the low-impedance line to the equipment. A suitable construction for such an inductance is illustrated in Chapter 6 (Fig 77).

## L and pi-section ATUs

The circuit given in Fig 89(a) is perhaps the simplest of all the multiband ATU designs. Capacitor C1 may be switched in as shown and this makes the circuit into a pi-section ATU (once called the 'Collins coupler'), but when it is switched out the circuit becomes an L-section system. The capacitor C1 does not need to be switched out, and can be left in circuit but set to its minimum capacitance.

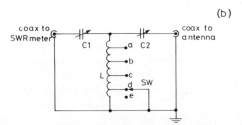

Fig 89. (a) An L-section or pi-matching unit. This is a versatile ATU which does not require a coupling coil. A single coil with adequate taps can be used to cover a wide range of frequencies. (b) A T-section ATU which is also a simple matching device. Its greatest disadvantage is that it is poor at rejecting high-frequency transmitter harmonics. Its construction is also complicated by the fact that the rotors and stators of C1 and C2 are not at earth potential

A wide range of impedance loads can be matched to 50ohms with this circuit and either a switched coil or a variable inductor (often called a 'roller coaster' or 'mangle') may be used for L. It will be found that if the antenna is at high impedance the capacitance of C1 will be greater than that of C2. A low-impedance input from the antenna requires the converse of this, and then C2 must be the largest capacitance.

In many ways this circuit resembles the previous ATU but the use of a considerable capacitance to ground via either C1 or C2 greatly reduces the chance of harmonic radiation. In effect this ATU is also a simple low-pass filter. It does not need a coupling coil and is also easy to adjust. Although the 'L' and 'pi' ATUs are good matching systems, their limitation is that they cannot match such a

wide range of impedances as some of the more sophisticated circuits that have been developed.

The component values for the pi-section ATU which will tune from 3.5 to 28MHz are as follows:

L = 15µH (20 turns of 14SWG, 3in. (76mm) diameter and 3.75in. (95mm) long. 10 turns are wound at four turns per inch (25mm) and 10 turns are at eight turns per inch. L is tapped every two turns)

C1 = 350pF (older-style valve receiver type)

C2 = 200pF (wide-spaced high-voltage type)

The coil may be wound on a former or air wound. If its diameter is reduced to 2in. (51mm), L will need 38 turns of 18SWG at six turns per inch.

## A T-network ATU

The T-network in Fig 89(b) has its inductor L in parallel with both the input and the output. This type of ATU is normally only used to match between a range of low impedances, and it is ideal when matching quite low values (say 10 to 50ohms) to standard 50ohm coaxial cables. The voltages across the capacitors C1 and C2 are not high, so wide-spaced transmitting types are not necessary. L can be tapped for multiband use and again no coupling coil is needed.

Unfortunately this circuit has a serious drawback: it has a very poor attenuation of transmitter harmonics and behaves as a high-pass filter, with L presenting a high shunt impedance. Nevertheless, it is useful and quite satisfactory when a good low-pass filter is used between the transmitter and the 'T' section. Some typical component values for this circuit are:

L = 22 turns of 16SWG enamelled wire 40mm (1 $^9/_{16}$in.) diameter, close wound. The taps down from the top of the coil (Fig 89(b)) are:

a = 8 turns (7-10MHz)
b = 5 turns (14MHz)
c = 5 turns (18-21MHz)
d = 3 turns (24-28MHz)
e = no tap – the complete coil is used on 3.5MHz.

C1 and C2 are both 160pF maximum capacitance broadcast types.

## The modified Z-match

The original Z-match (impedance-match) circuit was described as long ago as March 1948 when W1CJL's article appeared in the ARRL journal *QST*. Subsequent articles by other writers (including G6YR) appeared in the amateur radio press during the 'fifties but they were all based upon the original Z-match and were really intended

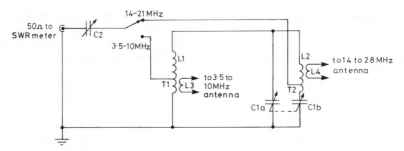

**Fig 90. The popular Z-match as modified by G5RV. This ATU is well suited to matching both tuned and untuned feedlines, and will cover all the HF bands (3.5 to 30MHz) without bandswitching**

to be coupled to the high-impedance end of valve PA anode circuits.

The connection of a 50ohm source via a variable capacitor to the 'top' of the Z-match multiband tuned circuit was poor engineering practice, and Louis Varney, G5RV, designed a much more satisfactory circuit which he described in the October 1985 issue of *Radio Communication*.

The basic G5RV modified Z-match is shown in Fig 90 and it can be seen that the 50ohm source is tapped down the main inductors L1 and L2 at T1 and T2. The variable input capacitor C2 is still included in Varney's design, and it is a part of the series-resonant tuned input circuit which, when correctly tuned, presents a 50ohm non-reactive load to the transmitter.

The Z-match uses the multiband tuner principle and can cover the 3.5 to 30MHz frequency range without coil changing. The inductance of the high-frequency coil L2 is small enough to be neglected at low frequencies, which means that the two sections of the split-stator tuning capacitor C1a and C1b are then in parallel and tune the LF inductor L1 over the 3.5 to 10MHz range. At the higher frequencies (14 to 28MHz) L1 is large enough to behave like an RF choke and it has little effect upon the L2 tuned circuit. On these frequencies L2 is tuned by the small variable capacitance provided by C1a and C1b which are in series.

The output coils L3 and L4 are both tightly coupled to L1 and L2 respectively, and are usually arranged to be of a larger diameter than these so they can actually be positioned over them. No output switching is shown in Fig 90, but it will be needed in a working model and should connect either L3 or L4 to a twin-wire feeder system. Other switching arrangements, which will allow the connection of unbalanced low-impedance coaxial feeders or instead connect the transmitter output directly to an antenna or dummy load, can be incorporated in the final practical design.

An SWR meter in the line between the transmitter and the Z-match will be needed when tuning this ATU. With C1a/C1b set to a median capacitance, the input capacitor C2 must then be adjusted for a minimum SWR reading. C1a/C1b should then be set to a different value of capacitance and the procedure repeated until the lowest possible SWR

is obtained. This process will have to be carried out on each band; a written note indicating the various capacitor settings will allow for rapid adjustment to the correct match on each band as needed.

The author's experiences with the Z-match revealed that it had a rather restricted range of impedance-matching capability, and when used with some antennas it was almost impossible to obtain a really low SWR on one or more bands. However, a great advantage of the circuit is that there is no need for coil changing or switching over a wide (8:1) frequency range.

Suggested component values for the modified Z-match (as given in *Radio Communication* October 1985, p776) are:

L1 =     10 turns at 40mm diameter close wound with 14SWG enamelled wire. The tap T1 is four turns from the earthy end.

L2 =     5 turns 14SWG enam., with the tap T2 made $1\frac{1}{2}$ turns from the centre of the coil towards C1b.

L3 =     8 turns 50mm diameter 14SWG enam. over the earthy end of L1.

L4 =     3 turns 14SWG enam. arranged over the tap T2 on L2.

C1a/C1b = split stator 250 + 250pF variable.

C2 =     500pF broadcast receiver type.

## The SPC matching circuit

For a number of years a modified version of the T-match circuit was used by many amateurs, particularly in the USA, and this was known as the 'ultimate transmatch'. Unfortunately, under some conditions of impedance transformation this circuit shows a high-pass response and does little to attenuate high-frequency harmonics. In the worst cases the attenuation of high-order harmonics might be as little as 3 to 6dB when using the 'ultimate transmatch', so to overcome this problem a new circuit was devised by W1FB.

This circuit is shown in Fig 91(a) and it is arranged so there will always be some capacitance in parallel with the

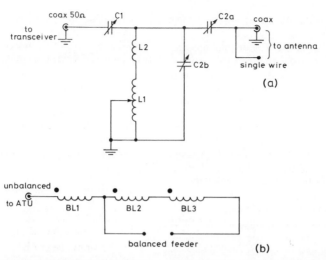

Fig 91. (a) The W1FB SPC matching circuit which can match a very wide range of impedances and also provide some considerable harmonic rejection. When using untuned twin wire feeders a balun is required. (b) A circuit for a trifilar air-wound 1:1 balun which is suitable for high-power work. At 100W or less a ferrite core can be used

inductor (L1 + L2). One half of a dual-section variable capacitor (C2b) tunes the inductance and its other section (C2a) is in series and connects the tuned circuit to the output connectors. It is the dual use of this capacitor that gives the name 'SPC' (series-parallel capacitance) to this ATU design.

The harmonic attenuation is good and the bandpass response is maintained when the load impedances range from less than 25ohms to more than 1000ohms. This is because there will always be a substantial capacitance in parallel with the inductor L1 + L2 and to earth.

A rotary inductor is frequently used for L1 and the smaller separate coil L2 is used at the high-frequency end of the ATU's range. This is a self-supporting inductor which must be positioned at right angles to L1. When constructing this design it should be borne in mind that the common and normally 'earthy' spindle between the two capacitor sections of C2a/C2b will not be connected to earth, but to the 'hot' end of the coil. This means that this component should be well insulated from the chassis and have an insulated extension spindle between it and the front panel.

As it is drawn in Fig 91, the SPC circuit can load into either a single wire or an unbalanced coaxial feeder, but when a 'flat' untuned feeder is used to feed the antenna this may terminate at a suitable (1:1 or 4:1) balun. The balun's unbalanced output will be to the ATU's antenna or coaxial cable connection. Some amateurs, as has been mentioned in an earlier chapter, use baluns in this way with tuned feedlines which have high standing waves along them. It is a practice not to be recommended and, although it may occasionally be possible to 'get away with it' on one or two bands when using a particular antenna and feeder length, on other frequencies there will be inevitable high reactances which cannot be tuned out. If the use of a tuned

feedline is contemplated it is best matched with a Z-match circuit or one of the circuits shown in Figs 87(b) and (c).

A suitable 1:1 balun will have 12 trifilar turns close wound on a 1in. diameter former. A ferrite core could saturate with high powers but can be used at 100W or less (see Fig 91(b) for details).

Construction of the SPC L1 is a rotary inductance (25µH minimum) or a tapped coil (every two turns) of 28 turns 14SWG wire, 2½in. (64mm) diameter over 3½in. (89mm). L2 is three turns of 10SWG (or ¼in. (6mm) wide copper strip), 1in. (25mm) diameter self-supporting over 1½in. (38mm). With an inductance of this specification (L1+L2) the SPC circuit will tune from 1.8 to 30MHz. C1 is 200pF and C2 is 200 + 200pF.

It is suggested that the wiring of an SPC matching unit should be made with thin flashing copper strips and that the stators of C2a/C2b are arranged so that they are well above the metal chassis. If this is not done the circuit might not resonate on 28MHz. If a metal cabinet or other enclosure is used it should be big enough to allow adequate spacing between it, the inductors and the variable capacitor stators (about twice the diameter of the coil L1). The author favours a tapped coil rather than a rotary inductor – once the correct tapping points have been found for each waveband, the bandswitching can be made very quickly. When a rotary inductance is used bandswitching can often be a lengthy operation!

## Matching Marconi antennas on 1.8MHz

The SPC circuit as described will tune down to 1.8MHz but many ATUs cannot be used below 3.5MHz. On 1.8MHz large values of inductance and capacitance are required so it is often advantageous to construct a separate matching unit for this band.

Fig 92(a) shows the typical use of a doublet antenna (often a G5RV type) with its open-wire feeders 'strapped' and the whole system tuned against ground. The total effective length of this antenna (half the top plus the feeder length) will usually be shorter than an quarter-wave at 1.8MHz, so there will be capacitive reactance at the feedpoint. This can be tuned out by using a simple series inductance L1. The coil may be of the rotary variable type or instead a fixed inductor which has a number of tap points along its length. L2 is a link winding and uses from three to six turns of insulated wire that is wound over the 'earthy' end of L1.

An RF current indicator (a hot-wire ammeter or a torch bulb!) in series with the antenna, together with an SWR meter in the coaxial cable line to the equipment, can be used to determine the required value of series inductance.

When an end-fed wire is longer than a quarter-wave (say, more than about 130ft long on 1.8MHz) it will exhibit inductive reactance. This can be tuned out with a variable capacitor C which is in series with the antenna (see Fig 92(b)). However, some series inductance can be used to provide an output coupling transformer L1/L2. The tuning of this arrangement is carried out in the same way as described for the circuit in Fig 92(a).

A more flexible ATU is shown in Fig 92(c). This can match almost any wire length, and it uses a tapped or variable inductor L1 together with a variable capacitor C between the bottom end of this inductor and earth. A flying lead (jumper) may be connected so that C can be shorted when it is not needed in circuit. Instead of a link winding this circuit has its 50ohm input connected directly over a few turns at the earthy end of L1. When the system is made properly resonant by the adjustment of L1 and/or C, the low-impedance feeder can then be tapped on to L1. A near-perfect match is possible if an SWR meter is used in the 50ohm feedline.

Suggested component values for the circuits in Figs 92(a), (b) and (c) are:

L1 = 30μH (20 turns 16SWG at 3in. (76mm) diameter over $3\frac{1}{2}$in. (89mm) tapped every two or three turns)

L2 = 3 to 6 turns of plastic insulated wire over the 'earthy' end of L1

C = 250pF variable (broadcast receiver type)

## An 'outboard' LC circuit

The author's considerable experience when using a variety of antennas and ATUs has taught him that there is no single matching unit that will give a 100% match on all bands to any type of antenna or feeder. It is very frustrating when an ATU which performs perfectly on five or six bands stub-

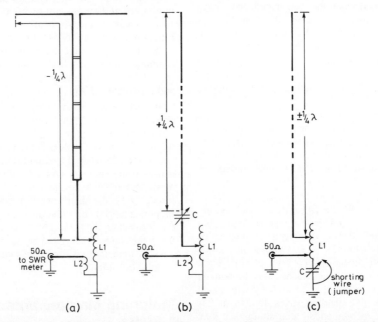

Fig 92. (a) Inductive loading a typical G5RV or similar antenna when used as a Marconi tuned against ground. The length of half the horizontal top of the antenna plus the length of the 'strapped' feeder should be less than a quarter-wave on 1.8MHz when this tuning method is employed. (b) If the wire to be tuned as a quarter-wave Marconi is actually longer than a quarter-wavelength, it will exhibit inductive reactance which can then be tuned out with a series capacitor C. The inductor L1 can still be employed as one winding of a transformer and provide coupling to the link winding L2. (c) This combination of L1 and C will allow almost any length of wire to be brought to quarter-wave resonance. The coaxial cable to the equipment is tapped on to L1 at a point which gives a match to its 50ohms impedance

bornly refuses to match a particular antenna on just one or two other bands.

A simple end-fed wire can present a wide range of feed impedances if it is used between 1.8 and 30MHz, and one is fortunate if it can be easily tuned and matched to the 50ohm transceiver input/output over this frequency range. If not, the author's answer to the problem is to have available an 'outboard' LC arrangement (see Fig 93) which can compensate for any unusual values of reactance or impedance at the feedpoint of single-wire antennas.

This circuit has a tapped inductance LR and a variable capacitor CR (with wide spacing), and it can be positioned close to the ATU antenna connection point. No switching is used, and to keep the unit simple and yet flexible three jumper leads with 'crocodile' clips at their ends can be employed to set up a variety of conditions.

As shown in Fig 93, a part of the inductance LR is used in series with the antenna and additionally CR is shown shunted to earth across the lead going to the main ATU. Using just CR without any series inductance is possible, or LR alone can be used. Sometimes just a series capacitor is needed and then the jumper lead at A is not connected. B will connect to A and the jumper shown joining to E can be connected to C. The inductor LR can be used in series with the antenna and CR can have one of its connections earthed and its jumper wire at B joined to point A.

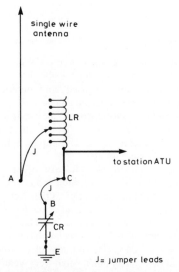

Fig 93. This is a universal 'outboard' matching unit which can be used between a single wire antenna and an ATU. In this way 'awkward' impedances and reactances can be translated to values that the station ATU can deal with

This arrangement has been found to work well and can compensate for unusual antenna reactances. It can be connected in a number of different configurations to allow the matching of even the most 'difficult' antennas. Suitable component values can be:

LR = 20µH (22 turns 14SWG at 2½in. (64mm) diameter over 2¾in. (70mm) and tapped every two turns)

CR = 200pF with wide plate spacing

## Commercially made ATUs

Thirty years ago there were few factory-built ATUs available in the UK and amateurs were obliged to construct their own. Now a very wide variety of matching units is available. Before an ATU purchase is contemplated, however, some care must be taken to ensure that it is going to be suitable for the antennas that are presently in use or are likely to be used in the future.

A first consideration must be the power-handling ability of an ATU. Many of them are only suitable for use with a basic transceiver and can therefore safely operate with transmitter output powers of about 150W at the most. If a linear amplifier is to be used the ATU must then be rated to somewhere between 500W and 2kW peak power; the variable-capacitor plate spacing seems to be the most important factor. It is the high voltage rating of the components used which make such ATUs expensive.

Another most important feature, which is not often explained satisfactorily in advertisements, is the ability of a commercial ATU to match a very wide range of impedances. Many of these units can only match a limited impedance range and are really only suitable for low-impedance input work. They have a working impedance of between 25 and 100ohms and are certainly not suitable for end-fed wire antennas, which on some bands may have an impedance of 1000ohms or more.

A good ATU (such as one based upon the SPC circuit) will match a very wide impedance range but many of the small matching units that have been designed to be used with particular 'black boxes' will only match antenna impedances (or feeder impedances) up to a maximum of 500ohms.

Most of the commercial ATUs now include several switching options which allow a variety of coaxial inputs, end-fed wires or even 'flat' untuned twin lines (using a balun) to be matched. They often also have a switched connection to an external dummy load and some have integral SWR meters. The more expensive models also have power output meters and they can be switched to several power level ranges. These ATUs can measure output powers from under 1W up to 2kW! Auto-tune ATUs are becoming more popular but many of these have a limited impedance-matching range.

## Other points to consider

Whatever type of ATU is used, its chassis must be the common earth point for the station, with all the equipments connecting to its earth terminal. Home-built ATUs do not need to be fully screened and a metal base and front panel are usually sufficient. Some amateurs use badly designed and inefficient matching units, and the author has found that in some cases this is because they were squashed into metal boxes which were much too small, with the coils almost touching the cabinet tops or sides. In other instances the wire used to wind the inductors was much too thin. A wire thickness of 14SWG or thicker should be used when winding ATU coils.

Many amateurs have experienced 'flash over' of the variable capacitors in their ATUs, especially on SSB peaks. This is either because the plate spacing of these capacitors is too small or the $Q$ of the tuned circuit is too high. A high-$Q$ circuit will not allow the proper transfer of power to the antenna, so there are high RF voltages built up which then cause 'flash over'.

Any sign of coil heating is a warning that the LC ratio or $Q$ is incorrect. A good and efficient ATU *never* has coil heating, for its throughput of energy is high. A loss of 3dB within an ATU means that half the transmitter power is contributing to heating and other losses.

## Odds and ends

### Wire and coils

The calculations which are needed when designing coils for specific values of inductance are quite involved, so Table 12 gives a useful selection of winding details for four different values.

**Table 12. Winding details for four values of inductance**

| Inductance (µH) | Wire (SWG) | Diameter (in.) | Length (in.) | Number of turns |
|---|---|---|---|---|
| 40 | 18 | 2$^1$/$_2$ | 2 | 28 |
| 40 | 14 | 2$^1$/$_2$ | 4$^1$/$_4$ | 34 |
| 20 | 18 | 2$^1$/$_2$ | 1$^1$/$_4$ | 17 |
| 20 | 14 | 2$^1$/$_2$ | 2$^3$/$_4$ | 22 |
| 8.6 | 16 | 2 | 2 | 16 |
| 8.6 | 14 | 2$^1$/$_2$ | 3 | 15 |
| 4.5 | 16 | 2 | 1$^1$/$_4$ | 10 |
| 4.5 | 14 | 2$^1$/$_2$ | 4 | 12 |

Here is the number of turns per inch of enamelled copper wire when it is closewound:

18SWG = 19.7 turns
16SWG = 14.8 turns
14SWG = 12.1 turns

The diameter of copper wire is as follows:

18SWG = 0.048in. or 1.22mm
16SWG = 0.064in. or 1.62mm
14SWG = 0.08in.  or 2.03mm

Our 18, 16 and 14SWG wires are the close equivalent of the USA 16, 14 and 12 wire sizes.

### Copper tubing

It is often an advantage to use really heavy gauge conductors when constructing coils. An example of this relates to antenna loading coils. In Chapter 1 ('Half-sized dipoles' and Fig 14) it is suggested that thick wire or tubing should be used to make the coils. A suitable tubing is the $^3$/$_{16}$in. (5mm) diameter copper which is sold as motor car brake fluid pipe and which can be obtained from most auto-repair shops or car accessory stores. It is sold in 30ft coils and is annealed. After the coils are wound they can be sprayed or painted with a good polyurethane varnish to prevent oxidation, especially if they are to be used out of doors.

### Insulators

The material used as an insulator over which antenna loading coils are wound (see Fig 14) can be a standard glass (Pyrex) item, or a length of a dense plastic called 'dia-polypenco-acetal'. This material is an acetal co-polymer and is manufactured by several companies under various brand names. One well-known name is 'Delrin' which is available from most plastics suppliers. This plastic has superb RF properties and great strength. A black Delrin/acetal is available and this has a greater resistance to the ultraviolet component in sunlight. Its electrical properties are identical to those of the standard white material.

### Weatherproofing

The modelling clay that is sold in the UK under the trade name 'Plasticene' makes a fine weatherproofing medium for any metallic hardware used with outdoor antenna installations. Nuts and bolts etc can be well covered with this plastic and oily substance and they will then remain fully protected against the corrosive effects of weathering for many years. Plasticene never completely hardens and does not deteriorate with age.

A mention has already been made of the silicone-rubber sealants that are available for kitchen and bathroom use. One writer in the radio amateur press has decried their use because of fears that the acetic acid which forms when this material changes from a liquid to a rubbery solid might attack metal parts. This kind of sealant has been used over several years at this QTH, and careful examinations of outdoor soldered connections which were weatherproofed with it at various times have never revealed any adverse effects upon the metals that were covered. The acetic acid is only present during the curing period and it finds its way to the surface and evaporates rapidly. The Dow Corning silicone-rubber sealant is a material that the author therefore feels he can endorse favourably.

### Halyards and guys

The use of old leaky buckets filled with small stones is

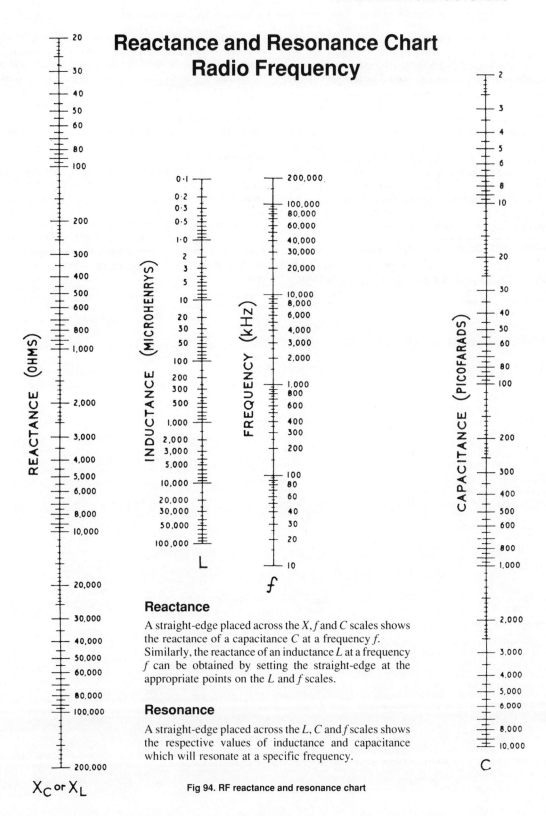

# Reactance and Resonance Chart
# Radio Frequency

REACTANCE (OHMS)

INDUCTANCE (MICROHENRYS)

FREQUENCY (kHz)

CAPACITANCE (PICOFARADS)

L

*f*

$X_C$ or $X_L$

C

### Reactance

A straight-edge placed across the $X$, $f$ and $C$ scales shows the reactance of a capacitance $C$ at a frequency $f$.
Similarly, the reactance of an inductance $L$ at a frequency $f$ can be obtained by setting the straight-edge at the appropriate points on the $L$ and $f$ scales.

### Resonance

A straight-edge placed across the $L$, $C$ and $f$ scales shows the respective values of inductance and capacitance which will resonate at a specific frequency.

**Fig 94. RF reactance and resonance chart**

often advocated for use as counterweights at the ground end of antenna halyards. When an antenna pulley is fixed to a tree or a branch, this weight will ride up or down in sympathy with the movements caused by the wind and so remove strains from the actual antenna wire.

Useful counterweights can be made with plastic one-gallon containers (similar in appearance to the 'jerrycans' used to hold motor fuel) in which some detergents and cooking oils are sold. These can be filled with just enough water to hold the antenna wire taut and then have their top caps screwed on tightly. A couple of similar weights in the author's garden have survived the attentions of the sun's ultraviolet radiation and severe freezing through four summers and winters without failure or cracking. Green containers  are used, which are almost invisible as they merge with the ground vegetation.

The expansion of water sports and sailing as popular recreations means that it is now possible to buy high-quality rot-proof ropes for halyards and guys, and also a variety of hardware items such as pulleys and turnbuckles etc, even in towns and cities that are far removed from the coast. Many of these items are for marine use and will not deteriorate for many years, even when exposed to the vagaries of our climate.

When erecting a new mast it is always a good idea to fix two pulley and halyard systems. Should one of them fail there will be no immediate need to take down and re-erect the mast, an operation that is not something to relish in mid-winter!

The synthetic materials used to make many ropes and cords will last longer if such ropes are soaked in black or other dark paint a few days before they are put to use. They will then have some protection from sunlight. It is now possible to obtain ropes etc which are already coloured and these are much better than the white varieties. Dark-coloured guy ropes are not so obvious to prying eyes and are less likely to offend the aesthetic feelings of neigh-bours!

# Postscript

This book has attempted to create an interest in the construction and use of simple wire antennas, and it is to be hoped that it may prove useful to both newcomers and also the more experienced followers of our hobby. For convenience, compactness and effectiveness a monoband rotary three-element or four-element beam antenna cannot be excelled, but many of the widely publicised and advertised tri-banders and mini-beams often have a disappointing performance. Several of the antennas described in this book can out-perform such compromise systems, and will be much cheaper.

It has been impossible to describe every possible variety of wire antenna; a noticeable omission is the trapped multiband dipole. The author has an abhorrence of most antennas which use resonant traps, and thinks that at best they are a poor substitute for full-length dipoles or multiband dipole systems. Most of the antennas that are described have been used and thoroughly air-tested at some time during the 42 years that the author been licensed, and the positive exclusion of trapped antennas is backed by a personal experience of their shortcomings.

Although exact copies can be made of the wire antennas that have been described, it is thought that enough extra basic information has been included to stimulate the reader to experiment with different versions. It is always impossible to please everyone, but perhaps each reader will find something of interest and usefulness somewhere within these pages.

# Index

# Some other RSGB publications...

## HF ANTENNAS FOR ALL LOCATIONS
This book explains the ''why'' as well as ''how'' of hf antennas, and takes a critical look at existing designs in the light of latest developments.

## AMATEUR RADIO AWARDS (third edition)
This new edition of Amateur Radio Awards gives details of major radio amateur awards throughout the world. Each award is listed in an easy to understand format giving all the information on how to achieve the award. An innovation for this edition is the provision of checklists so that the amateur can keep a record of progress. This book is essential reading for the avid award hunter and the dx chaser alike.

## AMATEUR RADIO OPERATING MANUAL
Covers the essential operating techniques required for most aspects of amateur radio including station organisation, and features a comprehensive set of operating aids.

## RADIO COMMUNICATION HANDBOOK
First published in 1938 and a favourite ever since, this large and comprehensive guide to the theory and practice of amateur radio takes the reader from first principles right through to such specialised fields as radio teleprinters, slow-scan television and amateur satellite communication.

## WORLD PREFIX MAP
This is a superb multi-coloured wall map measuring approximately 1200mm by 830mm. It shows amateur radio country prefixes worldwide, world time zones, IARU locator grid squares, and much more. A must for the shack wall of every radio amateur.

## RADIO DATA REFERENCE BOOK
Presents a wide range of useful reference material in convenient form, and without needless repetition of basic theory.

## RADIO SOCIETY OF GREAT BRITAIN
**Lambda House, Cranborne, Road, Potters Bar, Herts. EN6 3JE**

# RSGB — Representing Amateur Radio . . .

## Radio Communication

A magazine which covers a wide range of interests and which features the best and latest amateur radio news in its special Bulletin section. The Society's journal has acquired a world-wide reputation for its content. It strives to maintain its reputation as the best available and is now circulated, free of charge, to members in over 150 countries. The regular columns in the magazine cater for hf, vhf/uhf, microwave, swl, clubs, satellite, data, contests, and amateur TV. In addition to technical articles, the highly regarded Technical Topics feature caters for those wishing to keep themselves briefed on recent developments in technical matters. Major issues are discussed each month in the editorial column.

The "Last Word" is a lively feature in which members can put forward their views and opinions and be sure of receiving a wide audience. To keep members in touch with what's going on in the hobby, events diaries are published each month.

## Members' Advertisements

Subsidized advertisements for the equipment you wish to sell in the Society's monthly magazine, with short deadlines and large circulation.

## QSL Bureau

Members enjoy the use of the QSL Bureau free of charge for both outgoing and incoming cards. A leaflet is available from HQ.

Special Event Callsigns in the GB series are handled by RSGB. They give amateurs special facilities for displaying amateur radio to the general public. For details an application form, apply to the Membership Services Department at HQ. Please apply at least six weeks in advance.

## Specialized News Sheets

The weekly DX News-sheet for HF enthusiasts, the VHF/UHF Newsletter for VHF enthusiasts, the Microwave Newsletter for those operating above 1GHz and Connect International for packet radio enthusiasts. Details on request from the Circulation Department at HQ.

## Specialized Equipment Insurance

Insurance for your valuable equipment which has been arranged specially for members. The rates are very advantageous: details from HQ.

## Audio Visual Library

Films, audio and video tapes are available through one of the Society's Honorary Officers for all affiliated groups and clubs. Further details may be obtained either from the Honorary Officer (whose name can be found in Radio Communication) or from the Membership Services Department at HQ.

## Reciprocal Licensing Information

Always contact the Membership Services Department at HQ as early as you can if you plan to go abroad. Details are available for most countries on the RSGB computer data base.

## Government Liaison

One of the most vital features of the work of the RSGB is the ongoing liaison with the UK Licensing Authority - presently the Radiocommunications Division of the Department of Trade and Industry. Setting and maintaining the proper framework in which amateur radio can thrive and develop is essential to the well-being of amateur radio. The Society spares no effort in defence of amateur radio's most precious assets - the amateur bands.

## Beacons and Repeaters

The RSGB supports financially all repeaters and beacons which are looked after by the appropriate committee of the Society, ie, 1.8-30MHz by the HF Committee, 30-1000MHz (1GHz) by the VHF Committee and frequencies above 1GHz by the Microwave Committee. For repeaters, the Society's Repeater Management Group has played a major role. Society books such as the Amateur Radio operating Manual give further details, and computer based lists giving operational status can be obtained by post from HQ - see the price list for details of how to obtain these.

# . . . Representing You !

## Operating Awards

A wide range of operating awards are available via the responsible officers: their names can be found in the front pages of Radio Communication and in the Society's Members Handbook. The RSGB also publishes a book which gives details of most major awards.

### Contests (HF/VHF/Microwave)

The Society has two contest committees which carry out all work associated with the running of contests. The HF Contests Committee deals with contests below 30MHz, whilst events on frequencies above 30MHz are dealt with by the VHF Contests Committee.

## Morse Testing

In April 1986 the Society took over responsibility for morse testing of radio amateurs in the UK. If you wish to take a morse test write direct to RSGB HQ (Morse tests - BR) for an application form.

### Slow Morse

Many volunteers all over the country give up their time to send slow morse over the air to those who are preparing for the 12 words per minute morse test. You can find the schedule in Radio Communication, or the Members Handbook. The Society also produces morse instruction tapes.

## RSGB Books

The Society publishes a range of books for the radio amateur and imports many others. The price list and ordering details can usually be found at the back of Radio Communication. RSGB members are entitled to a 15% discount on all books purchased from the Society. This discount can offset the cost of membership.

## Propagation

The Society's Propagation Studies Committee is highly respected - both within the amateur community and professionally - for its work. Predictions are given in the weekly GB2RS news bulletins, the Society's monthly magazine Radio Communication.

## Technical Advice

Although the role of the Society's Technical and Publications Committee is largely to vet material intended for publication, its members and HQ staff are always willing to help with any technical matters.

### EMC Advice

Breakthrough in domestic entertainment equipment can be a difficult problem to solve as well as having licensing implications. The Society's EMC Committee is able to offer practical assistance in many cases. The Society also publishes a special book to assist you. Additional advice can be obtained from the EMC Committee Chairman via RSGB HQ.

## Planning Permission

There is a special booklet and expert help available to members seeking assistance with planning matters.

## GB2RS

A special radio news bulletin transmitted each week and aimed especially at the UK radio amateur and short wave listener. The script is prepared each week by the Society's HQ staff, and items of news can be left on the special telephone answering machine on 0707 59260. The transmission schedule for GB2RS is printed regularly in Radio Communication, or it can be obtained via the Membership Services Department at HQ. It also appears in the Members Handbook. The GB2RS bulletin is also sent out over the packet radio network.

## Raynet (Radio Amateur Emergency Network)

Several thousand radio amateurs give up their free time to help with local, national and sometimes international emergencies. There is also ample opportunity to practice communication and liaison skills at non-emergency events, such as county shows and charity walks, as a service to the people. For more information or details of how to join, contact the Membership Services Department at HQ.

# Notes